VOLUME EIGHTY-TWO

ADVANCES IN
CARBOHYDRATE CHEMISTRY AND BIOCHEMISTRY

Special Volume in Memory of Hidetoshi Yamada Part 2

VOLUME EIGHTY-TWO

Advances in
CARBOHYDRATE CHEMISTRY AND BIOCHEMISTRY
Special Volume in Memory of Hidetoshi Yamada Part 2

Editor

DAVID C. BAKER
*University of Tennessee,
Knoxville, TN, USA*

ELSEVIER

ACADEMIC PRESS
An imprint of Elsevier

Academic Press is an imprint of Elsevier
50 Hampshire Street, 5th Floor, Cambridge, MA 02139, United States
525 B Street, Suite 1650, San Diego, CA 92101, United States
The Boulevard, Langford Lane, Kidlington, Oxford OX5 1GB, United Kingdom
125 London Wall, London, EC2Y 5AS, United Kingdom

First edition 2022

Notices
Knowledge and best practice in this field are constantly changing. As new research and experience
broaden our understanding, changes in research methods, professional practices, or medical
treatment may become necessary.

Practitioners and researchers must always rely on their own experience and knowledge in evaluating
and using any information, methods, compounds, or experiments described herein. In using such
information or methods they should be mindful of their own safety and the safety of others, including
parties for whom they have a professional responsibility.

To the fullest extent of the law, neither the Publisher nor the authors, contributors, or editors, assume
any liability for any injury and/or damage to persons or property as a matter of products liability,
negligence or otherwise, or from any use or operation of any methods, products, instructions, or ideas
contained in the material herein.

ISBN: 978-0-323-98597-0
ISSN: 0065-2318

For information on all Academic Press publications
visit our website at https://www.elsevier.com/books-and-journals

Publisher: Zoe Kruze
Acquisitions Editor: Jason Mitchell
Developmental Editor: Nadia Santos
Production Project Manager:
   Vijayaraj Purushothaman
Cover Designer: Mark Rogers

Typeset by STRAIVE, India

Transferred to Digital Printing 2022

Working together
to grow libraries in
developing countries

www.elsevier.com • www.bookaid.org

Contents

Contributors

Go Hirai
Graduate School of Pharmaceutical Sciences, Kyushu University, Fukuoka City, Japan

Kyohei Muguruma
Department of Chemical Science and Engineering, School of Materials and Chemical Technology, Tokyo Institute of Technology, Tokyo, Japan

Toshiki Nokami
Department of Chemistry and Biotechnology, Tottori University, Tottori, Japan

Kaname Sasaki
Department of Chemistry, Toho University, Funabashi, Japan

Daisuke Takahashi
Department of Applied Chemistry, Faculty of Science and Technology, Keio University, Yokohama, Japan

Katsunori Tanaka
Department of Chemical Science and Engineering, School of Materials and Chemical Technology, Tokyo Institute of Technology, Tokyo; and Biofunctional Synthetic Chemistry Laboratory, RIKEN, Saitama, Japan

Kazunobu Toshima
Department of Applied Chemistry, Faculty of Science and Technology, Keio University, Yokohama, Japan

Nanako Uesaki
Department of Chemistry, Toho University, Funabashi, Japan

Kenshiro Yamada
Department of Chemical Science and Engineering, School of Materials and Chemical Technology, Tokyo Institute of Technology, Tokyo, Japan

Preface

Volume 82 continues with Part 2 of contributed papers dedicated to the memory of Prof. Hidetoshi Yamada.[a] This volume consists of five chapters that cover synthetic chemistry, including a chapter on the in vivo synthesis of bioactive compounds and associated mechanistic studies in the synthesis of a variety of biologically important carbohydrates.

In Chapter 1, entitled "Toward a one-pot, selective synthesis of cyclic oligosaccharides," Toshiki Nokami details the chemistry involved in a novel, one-pot electrochemical polyglycosylation–isomerization–cyclization process for the synthesis of cyclic oligosaccharides. The remarkable process is particularly applicable to the synthesis of unnatural cyclic oligosaccharides for which few methods are available.

Katsunori Tanaka and colleagues Kenshiro Yamada and Kyohei Muguruma in Chapter 2 explore the exciting topic of "Therapeutic in vivo synthetic chemistry using an artificial metalloenzyme with glycosylated human serum albumin." By addressing the concept of "therapeutic in vivo synthetic chemistry," which makes use of chemical synthesis carried out in living organisms for the diagnosis and treatment of diseases, these researchers have devised novel, targeted therapies for cancer using a combination of glycan-modified human serum albumin and nonnatural metal catalysts that are selective for targeting specific glycans. The authors correctly suggest that such approaches hold promise for the treatment of a myriad of disease states.

Chapter 3, entitled "Pseudo glycosides with a C-glycoside linkage," by Go Hirai covers the chemistry and biological activities of a most interesting set of compounds: pseudo-glycoconjugates, specifically those in which the O in the sialyl linkage is replaced by C. In compounds related to GM3, stereoselective chemical approaches to stable, bioactive compounds are evaluated, ultimately leading to a CHF-sialoside-GM3-type compound. The inclusion of an F atom in the linkage provides conformational control and improved biological activity. A bold direct C-glycosylation method using atom transfer radical coupling in the syntheses of pseudo-isomaltose and pseudo-KRN7000 is also introduced.

---

[a] For Part 1, see *Advances in Carbohydrate Chemistry and Biochemistry*. Baker, D. C., Ed. Academic Press/Elsevier: Oxford, 2022; Vol. 81.

Chapter 4, entitled "Boron-mediated aglycon delivery (BMAD) for the stereoselective synthesis of 1,2-*cis*-glycosides," by Daisuke Takahashi and Kazunobu Toshima addresses the challenges in the synthesis of 1,2-*cis*-glycosides, a class of biologically important oligosaccharides. Unlike their 1,2-*trans*-counterparts, these 1,2-*cis*-glycosides cannot rely on 1,2-*O*-acyl neighboring group participation to provide regio- and stereoselectivity in their synthesis. The authors review 1,2-*cis*-glycosidation methodologies involving boron-mediated reagents to solve these problems, the most impressive of which is their 1,2-*cis*-stereoselective glycosylation method called boron-mediated aglycon delivery. Variations on this theme allow for the direct *cis*-glycosylation of unprotected sugars, which holds promise for future developments in simplified approaches to these all-important compounds.

A long-sought-after goal in carbohydrate chemistry is to be able to synthesize specific anomers of glycosides at will and with a high degree of selectivity. Approaches to these problems are reviewed by Kaname Sasaki and Nanako Uesaki in Chapter 5, entitled "Conformationally restricted donors for stereoselective glycosylation." The authors discuss routes via control of glycosidation through control of the conformation of the glycosyl donor through manipulating protecting groups that affect their conformations and hence the reactivity and stereoelectronics involved in the glycosidation reaction. The collected examples are impressive and quite realistically illustrate workable solutions to the anomer control problem. Mechanistic arguments provide the basis for rationalizations for design of complex glycosides.

This volume concludes an impressive collection of papers dedicated to the memory of Prof. Hidetoshi Yamada, which includes a mix of synthetic chemistry, in vivo synthesis, and biological aspects of a variety of glycans that should attract the interest of a broad spectrum of carbohydrate researchers.

DAVID C. BAKER
Knoxville, TN 37919
October 2022

CHAPTER ONE

Towards one-pot selective synthesis of cyclic oligosaccharides ☆

Toshiki Nokami*

Department of Chemistry and Biotechnology, Tottori University, Tottori, Japan
*Corresponding author: e-mail address: tnokami@tottori-u.ac.jp

Contents

Abbreviations

CD	cyclodextrin [a cyclic oligosaccharide specifically a cyclomaltooligosaccharide (IUPAC)]
DMDO	dimethyldioxirane
(ePIC)	electrochemical polyglycosylation–isomerization–cyclization
NIS	N-iodosuccinimide
TfOH	triflic acid (trifluoromethanesulfonic acid)

1. Introduction

Professor Hidetoshi Yamada (Hidetoshi) and I met for the first time at the annual meeting of the Chemical Society of Japan. Since then we have been friends for more than 15 years. At that time, he had started developing a bridging protecting group that was useful for constructing β-glycosidic linkages in the absence of neighboring-group participation.[1] In April 2019,

☆ Dedicated to the memory of late Professor Hidetoshi Yamada.

Advances in Carbohydrate Chemistry and Biochemistry, Volume 82
ISSN 0065-2318
https://doi.org/10.1016/bs.accb.2022.10.004

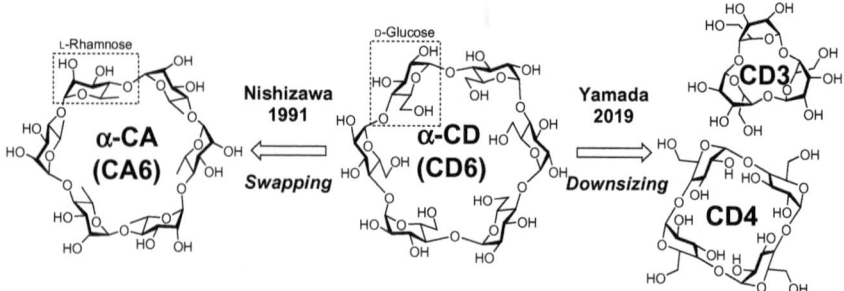

Fig. 1 Structures of α-cyclodextrin (cyclomaltohexaose) and other cyclic oligosaccharides.

he reported the first synthesis of the smallest of the cyclodextrins with three and four glucose repeating units (CD3 and CD4) (Fig. 1, right).[2] This work is a milestone in carbohydrate research, and its wide application can be expected. In August of the same year, Hidetoshi and I were talking about his research career on the train bound for San Diego. Then I recognized that Hidetoshi's journey to the unnatural cyclic oligosaccharides had already started many years ago. After completion of his master's degree studies at Osaka City University, he joined the Nishizawa group in Tokushima Bunri University as a faculty member and obtained his Ph.D. degree there. The Nishizawa group was dedicated to the isolation and synthesis of natural products based on their own synthetic methodologies; however, the synthetic study of an unnatural cyclic oligosaccharide called "Cycloawaodorin" was also a famous work of the group (Fig. 1, left).[3,4] Cycloawaodorin, named after the traditional summer dance "Awaodori" in Tokushima prefecture, is an unnatural cyclic oligosaccharide composed of L-rhamnose as the repeating unit. This oligosaccharide is one of my favorite molecules because of its structure and name.

In October 2019, we invited Prof. Yamada to the regional symposium of the Society of Synthetic Organic Chemistry, Japan at Tottori University, and he talked about his research, dreams, and hopes for the younger generations of chemists (Fig. 2). After the symposium we discussed how the chemical synthesis of cyclic oligosaccharides can be improved. He had already achieved the synthesis of the smallest cyclodextrin (CD3); however, many steps were required to complete the total synthesis. Therefore, we believed that the controlled polymerization of a monosaccharide building block was an ideal approach for the synthesis (Fig. 3). Of course, it must be very challenging to control the size and stereochemistry of cyclic

Fig. 2 Hidetoshi discussing in his lecture at the regional symposium of the Society of Synthetic Organic Chemistry, Japan at Tottori University, October 2019.

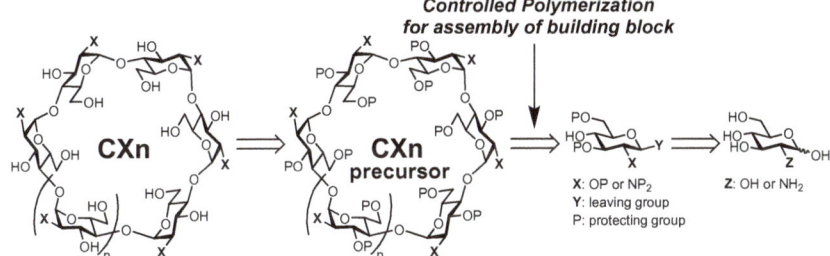

Fig. 3 Controlled polymerization in the retrosynthetic analysis of cyclic oligosaccharides **CXn**.

oligosaccharides by glycosylation polymerization (polyglycosylation). He had many ideas about the selective synthesis of cyclic oligosaccharides from monosaccharide building blocks under controlled polyglycosylation conditions. Although his journey suddenly ended several weeks later, I feel that we are still walking together.

 ## 2. Electrochemical conversion of linear oligosaccharides to cyclic oligosaccharides

There are two major problems in the synthesis of cyclic oligosaccharides. The first problem is the formation of linear oligosaccharides as precursors of the cyclic oligosaccharides, and the second problem is the

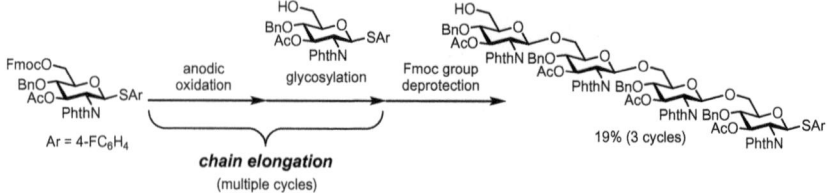

Fig. 4 Synthesis of the oligosaccharides via automated electrochemical assembly.

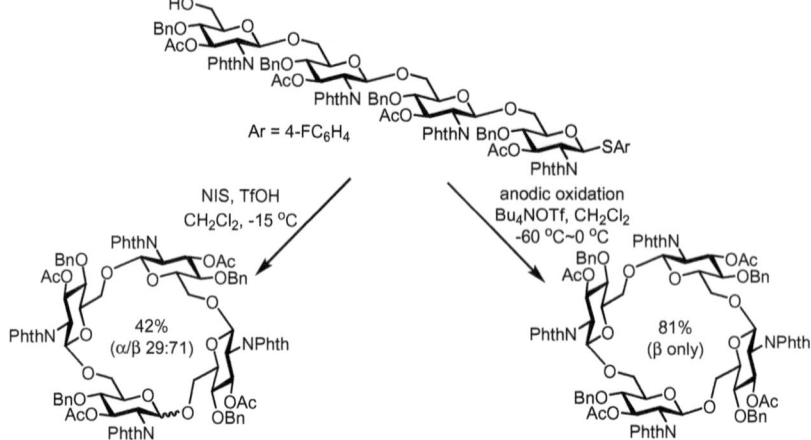

Fig. 5 Comparison of conventional chemical glycosylation and electrochemical glycosylation.

conversion of those linear oligosaccharides to the corresponding cyclic oligosaccharides. We had already developed an automated electrochemical assembly, which is an electrochemical, one-pot iterative synthesis of oligosaccharides.[5] Thus, we envisioned that linear oligosaccharides obtained by this method must be useful as precursors of cyclic oligosaccharides. A thioglycoside equipped with a 9-fluorenylmethoxycarbonyl (Fmoc) group as a temporary protecting group was used for automated electrochemical assembly to synthesize the precursors of cyclic oligosaccharides in one pot (Fig. 4).[6] Although yields of the linear oligosaccharides were moderate, formation of multiple glycosidic bonds and deprotection of the Fmoc group were carried out in an automated manner.

Next, the linear tetrasaccharides were electrochemically converted to the corresponding cyclic tetrasaccharides by intramolecular glycosylation in high yields and in perfect β-stereoselectivity (Fig. 5, right). By contrast, conventional chemical activation using *N*-iodosuccimide (NIS) and triflic acid

(TfOH) gave the corresponding cyclic oligosaccharides in the form of a mixture of isomers (Fig. 5, left).[7,8] Indeed, electrochemical methods contributed to the rapid synthesis of cyclic oligosaccharides; however, the synthesis is still time-consuming, and purification of the linear oligosaccharides is also tedious.

3. Electrochemical polyglycosylation to prepare chitin oligosaccharides

Polyglycosylation is a practical method to prepare linear or cyclic oligosaccharides with a single repeating structure.[9,10] Therefore, we started investigating the polyglycosylation of thioglycosides under electrochemical conditions. The building block with a protecting-group-free hydroxyl group at the C-4 position (4-OH) afforded the protected precursors of chitin oligosaccharides up to the hexasaccharide level (Fig. 6).[11] Chain length of the linear oligosaccharides depended on the structures of the thioglycoside building blocks and reaction parameters such as the amount of electricity and temperature. For example, the anomeric leaving group influenced the chain length of the linear oligosaccharides. The thioglycoside building blocks with the anomeric 4-fluorophenyl group afforded oligosaccharides in the highest total yields, and the longest oligosaccharide was hexasaccharide ($n=6$).

To the contrary, the building block with a protecting-group-free hydroxyl group at the C-6 position (6-OH) afforded the 1,6-anhydrosugar and the cyclic disaccharide as major products (Fig. 7).[12] This result suggested

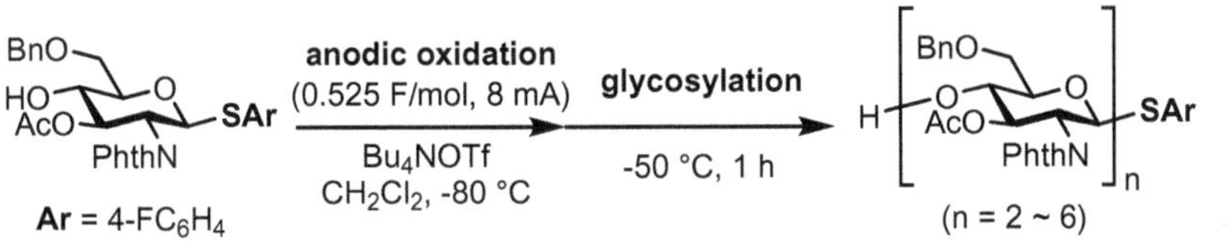

Fig. 6 Electrochemical polyglycosylation of a thioglycoside building block derived from glucosamine.

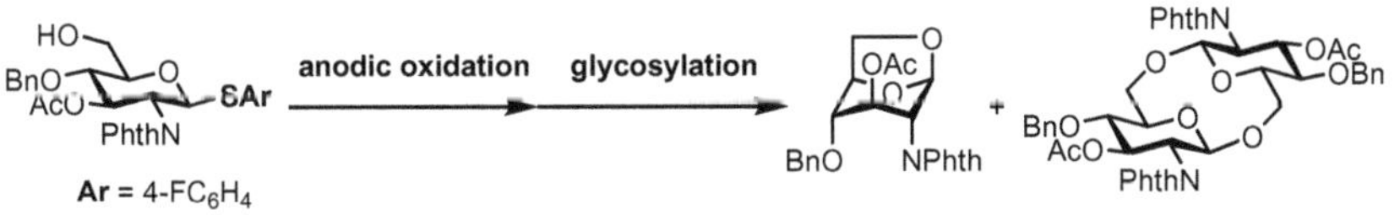

Fig. 7 Products of the thioglycoside building block with protecting-group-free 6-OH under the electrochemical polyglycosylation conditions.

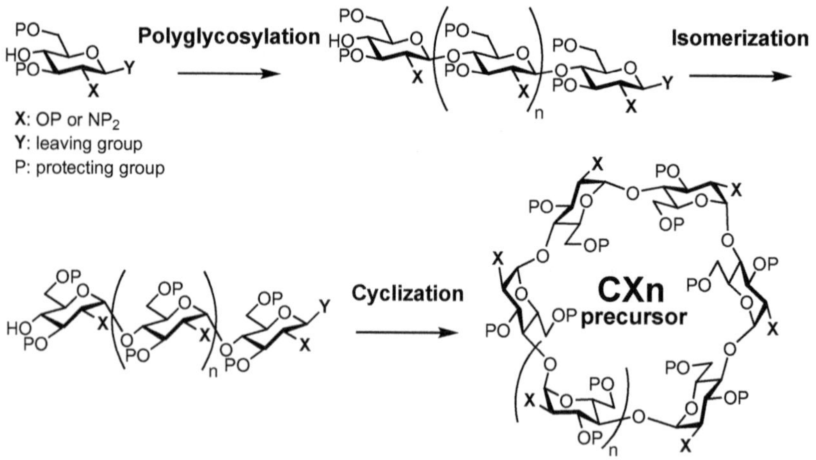

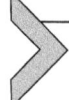

Fig. 8 Synthetic procedure of the precursor of cyclic oligosaccharides via isomerization.

that intramolecular glycosylation of monosaccharides and disaccharides occurred predominantly, and further chain elongation of the linear oligosaccharides was prevented. Therefore, it was crucial to prevent intramolecular glycosylation of short oligosaccharides to obtain larger cyclic oligosaccharides. Although longer oligosaccharides up to hexasaccharide were obtained using thioglycoside building blocks with free 4–OH, it was impossible to cyclize the oligosaccharides with β-1,4-glycosidic linkages because of their linear structures.

These results inspired us to develop a novel synthetic procedure of cyclic oligosaccharides with α-1,4-glycosidic linkages via one-pot isomerization (Fig. 8) The linear oligosaccharides with β-1,4-glycosidic linkages had already been synthesized under the electrochemical polyglycosylation conditions. Therefore, we envisioned that the sequential isomerization and cyclization of the linear oligosaccharides might afford cyclic oligosaccharides with α-1,4-D-glycosidic linkages.

4. One-pot polyglycosylation–isomerization–cyclization process

Based on the results of polyglycosylation, we designed the one-pot electrochemical process to produce "cyclokasaodorin," which is an N-acetyl-D-glucosamine analogue of α-cyclodextrin (Fig. 9).[13] When we compared the structures of α-cyclodextrin and cyclokasaodorin, the only

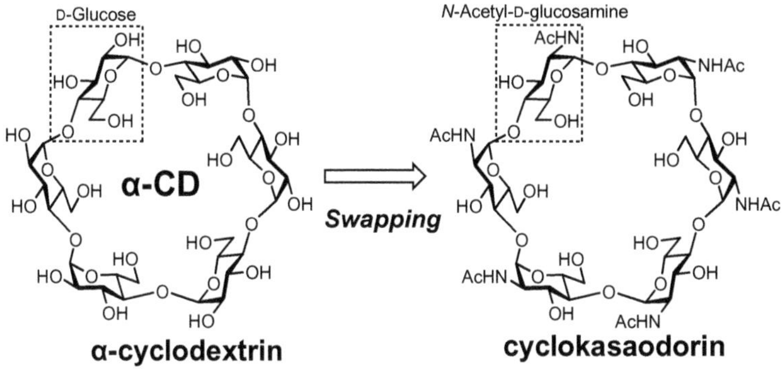

Fig. 9 Structural comparison of α-cyclodextrin and cyclokasaodorin.

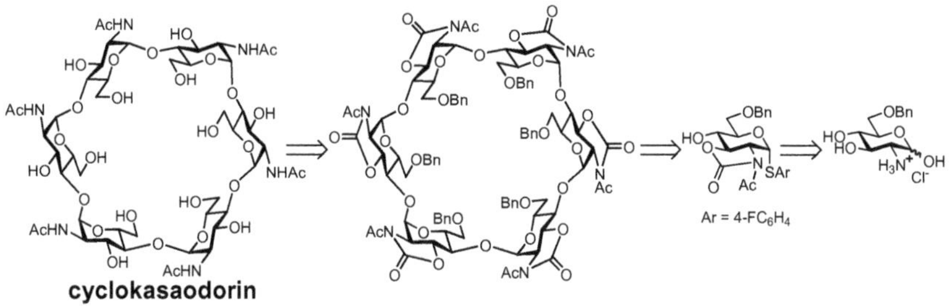

Fig. 10 Retrosynthesis of cyclokasaodorin.

difference was the substituents at the C-2 positions of the pyran ring. Although complete and site-selective conversion of the hydroxyl groups of α-cyclodextrin to amino groups has already been reported, it was limited to the hydroxy group at the C-6 positions.[14] Therefore, we planned to synthesize cyclokasaodorin via electrochemical assembly of the monosaccharide building blocks (Fig. 10). To achieve the synthesis of cyclokasaodorin, a thioglycoside equipped with a 2,3-oxazolidinone protecting group was chosen because the protecting group assisted the isomerization of β-glycosidic linkages to α-glycosidic linkages under electrochemical conditions.[15] Then, we developed the one-pot electrochemical polyglycosylation–isomerization–cyclization (ePIC) process to prepare the protected precursor of cyclokasaodorin efficiently (Fig. 11). The precursors of hexasaccharide (n = 4) and heptasaccharide (n = 5) were obtained in 6.2% and 5.5% yields, respectively. According to the reported procedures,[16,17]

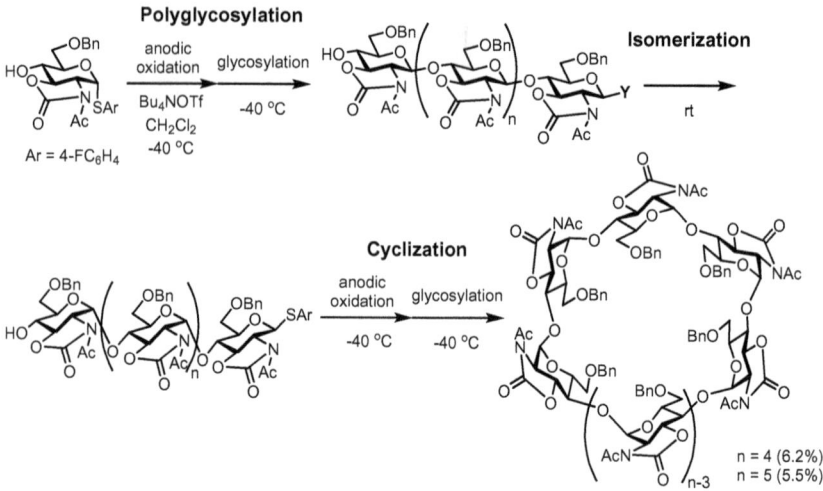

Fig. 11 The one-pot electrochemical polyglycosylation-isomerization-cyclization (*e*PIC) process.

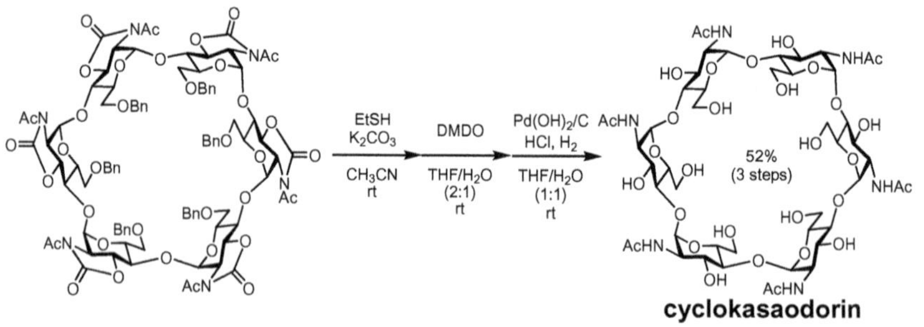

Fig. 12 Deprotection procedure for the total synthesis of cyclokasaodorin.

deprotection of the precursor of the cyclic hexasaccharide was carried out and cyclokasaodorin, which is cyclic hexasaccharide of α-1,4-linked *N*-acetylglucosamine, was obtained in 52% yield (three steps) (Fig. 12).

This synthetic study of cyclokasaodorin pays homage to the unique work of cycloawaodorin by Nishizawa and coworkers, including Hidetoshi Yamada.[3,4] Although "Kasaodori" has not yet become as famous as "Awaodori" in Tokushima, Kasaodori is a traditional summer dance in the eastern part of Tottori prefecture and was certified as the world's largest umbrella dancing in 2014 (Fig. 13).

Cyclokasaodorin

Fig. 13 Umbrella dancing on the main street "Wakasa-Kaido" of Tottori city. Image and picture of umbrella dancing ©Tottori Shan Shan Festival Promotion Association.

5. Conclusions

In this article the synthesis of cyclic oligosaccharides using electro-chemical methods is introduced. Although automated electrochemical assembly has been used to prepare linear oligosaccharides, which are precursors of cyclic oligosaccharides, this was a very conservative approach. Polyglycosylation is an interesting alternative; however, it is not selective. Finally, the *e*PIC approach was developed to prepare the precursors of unnatural cyclic oligosaccharides in one pot. It is a relatively challenging approach to synthesize cyclic oligosaccharides, and further optimization of the reaction will be required to provide a variety of cyclic oligosaccharides in better yields. During my preparation of this article, I was remembering Hidetochi's career, work, and words. He believed and expected the younger generation to do something new and interesting. In that vein, I would like to dedicate this article to his memory.

References

1. Okada, Y.; Asakura, N.; Bando, M.; Ashikaga, Y.; Yamada, H. Completely β-Selective Glycosylation Using 3,6-*O*-(*o*-xylylene)-Bridged Axial-Rich Glucosyl Fluoride. *J. Am. Chem. Soc.* **2012**, *134*, 6940–6943.
2. Ikuta, D.; Hirata, Y.; Wakamori, S.; Shimada, H.; Tomabechi, Y.; Kawasaki, Y.; Ikeuchi, K.; Hagimori, T.; Matsumoto, S.; Yamada, H. Conformationally Supple Glucose Monomers Enable Synthesis of the Smallest Cyclodextrins. *Science* **2019**, *364*, 674–677.

3. Nishizawa, M.; Imagawa, H.; Kan, Y.; Yamada, H. Total Synthesis of cyclo-L-Rhamnohexaose by a Stereoselective Thermal Glycosylation. *Tetrahedron Lett.* **1991**, *32*, 5551–5554.
4. Nishizawa, M.; Imagawa, H.; Kubo, K.; Kan, Y.; Yamada, H. Improved Synthesis of α-Cyclowaodorin. *Synlett* **1992**, 447–448.
5. Nokami, T.; Hayashi, R.; Saigusa, Y.; Shimizu, A.; Liu, C.-Y.; Mong, K.-K. T.; Yoshida, J.-I. Automated Solution-Phase Synthesis of Oligosaccharides via Iterative Electrochemical Assembly of Thioglycosides. *Org. Lett.* **2013**, *15*, 4520–4523.
6. Manmode, S.; Tanabe, S.; Yamamoto, T.; Sasaki, N.; Nokami, T.; Itoh, T. Electrochemical Glycosylation as an Enabling Tool for the Stereoselective Synthesis of Cyclic Oligosaccharides. *ChemistryOpen* **2019**, *8*, 869–872.
7. Gening, M. L.; Titov, D. V.; Grachev, A. A.; Gerbst, A. G.; Yudina, O. N.; Shashkov, A. S.; Chizhov, A. O.; Tsvetkov, Y. E.; Nifantiev, N. E. Synthesis, NMR, and Conformational Studies of Cyclic Oligo-(1→6)-β-D-Glucosamines. *Eur. J. Org. Chem.* **2010**, *2465–2475*.
8. Titov, D. V.; Gening, M. L.; Gerbst, A. G.; Chizhov, A. O.; Tsvetkov, Y. E.; Nifantiev, N. E. *Carbohydr. Res.* **2013**, *381*, 161–178.
9. Hashimoto, H.; Abe, Y.; Yoshimura, J. Synthesis of Chitooligosaccharide Derivatives by Condensation Polymerization. *J. Carbohydr. Chem.* **1989**, *8*, 307–311.
10. Someya, H.; Seki, T.; Ishigami, G.; Itoh, T.; Saga, Y.; Yamada, Y.; Aoki, S. One-Pot Synthesis of Cyclic Oligosaccharides by the Polyglycosylation of Monothioglycosides. *Carbohydr. Res.* **2020**, *487*, 107888.
11. Rahman, M. A.; Kuroda, K.; Endo, H.; et al. Synthesis of Protected Precursors of Chitin Oligosaccharides by Electrochemical Polyglycosylation of Thioglycosides. *Beilstein J. Org. Chem.* **2022**, *18*, 1133–1139.
12. Rahman, M. A.; Yamamoto, T.; Nokami, T. *unpublished results*; 2022.
13. Endo, H.; Ochi, M.; Rahman, M. A.; Hamada, T.; Kawano, T.; Nokami, T. Synthesis of Cyclic α-1,4-Oligo-N-acetyllglucosamine 'cyclokasaodorin' via a One-Pot Electrochemical Polyglycosylation-isomerization-Cyclization Process. *Chem. Commun. (Cambridge)* **2022**, *58*, 7948–7951.
14. Madhavan, N.; Robert, E. C.; Gin, M. S. A Highly Active Anion-Selective Aminocyclodextrin Ion Channel. *Angew. Chem., Int. Ed.* **2005**, *44*, 7584–7587.
15. Nokami, T.; Shibuya, A.; Saigusa, Y.; Manabe, S.; Ito, Y.; Yoshida, J.-I. Electrochemical Generation of 2,3-Oxazolidinone Glycosyl Triflates as an Intermediate for Stereoselective Glycosylation. *Beilstein J. Org. Chem.* **2012**, *8*, 456–460.
16. Taber, D. F.; DeMatteo, P. W.; Hassan, R. A. Simplified Preparation of Dimethyldioxirane (DMDO). *Org. Synth.* **2013**, *90*, 350–357.
17. Koinuma, R.; Tohda, K.; Aoyagi, T.; Tanaka, H. Chemical Synthesis of α(2,8) Octasialosides, the Minimum Structure of Polysialic Acids. *Chem. Commun. (Cambridge)* **2020**, *56*, 12981–12984.

CHAPTER TWO

Therapeutic in vivo synthetic chemistry using an artificial metalloenzyme with glycosylated human serum albumin ☆

Kenshiro Yamada[a], Kyohei Muguruma[a], and Katsunori Tanaka[a,b],*

[a]Department of Chemical Science and Engineering, School of Materials and Chemical Technology, Tokyo Institute of Technology, Tokyo, Japan
[b]Biofunctional Synthetic Chemistry Laboratory, RIKEN, Saitama, Japan
*Corresponding author: e-mail address: kotzenori@riken.jp

Contents

Abbreviations

2,3-Sia   2,3-sialylated glycan
2,6-Sia   2,6-sialylated glycan
ArM   artificial metalloenzyme
Gal   D-galactose
GArM   glycan modified artificial metalloenzyme;
GlcNAc   2-acetamido-2-deoxy-D-glucose (N-acetylglucosamine)
glycoHSA   glycan-modifed human serum albumin

☆In memory of Professor Hidetoshi Yamada.

Advances in Carbohydrate Chemistry and Biochemistry, Volume 82
ISSN 0065-2318
https://doi.org/10.1016/bs.accb.2022.10.001

GSH	glutathione
Man	D–mannose
SeCT	selective cell tagging
TON	turnover number

1. Introduction

Drug candidates are often eliminated from the development process regardless of pharmacological efficacy due to features such as toxicity, poor solubility, and instability. To change their unfavorable properties, the strategy of developing a prodrug can be used, whereby the functional group of the drug molecule is masked as an inactivated form.[1] Subsequently, the masking group is removed, allowing drug release through reactions existing in living systems such as the hydrolysis or reduction mediated by over-expressed enzymes at the target site (Fig. 1A).[2]

In comparison with these classic prodrug activations using natural reactions in vivo, there are methods available that use new-to-nature chemical reactions to trigger drug release (Fig. 1C). As represented by click-to-release,[3–5] controlled drug release using abiotic biorthogonal reactions has also been attracting attention because the abiotic structures are less susceptible to biological systems, thereby enhancing the specificity of drug release without unexpected drug activation. In addition, some researchers have developed biocompatible metal catalysts[6] that induce drug release by de-caging masking groups and also through abiotic chemical synthesis of organic small molecules in vivo, as if in a glass flask. For example, cyclic drug molecules are synthesized by intramolecular cyclization from chain-like precursors possessing greatly altered pharmacological and physical properties from the parental drugs (Fig. 1D). Alternatively, the metal catalyst can be applied to chemical tagging-mediated disease therapy [selective cell tagging (SeCT) therapy] by tagging the cellular surface of functional molecules, which allows it to exert its function, such as cytotoxic activity, only on the tagged cell (Fig. 1E). We defined the term "therapeutic in vivo synthetic chemistry" to include these concepts and to refer to chemical synthesis in living systems using new-to-nature reactions for the treatment or diagnosis of diseases.[7] This approach is anticipated to be a new modality for disease treatment.

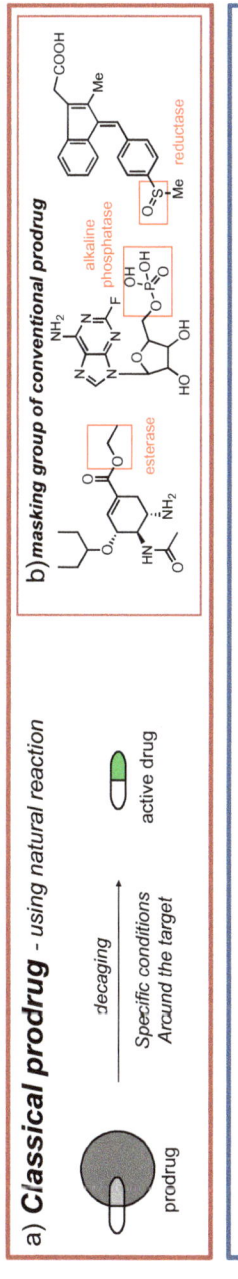

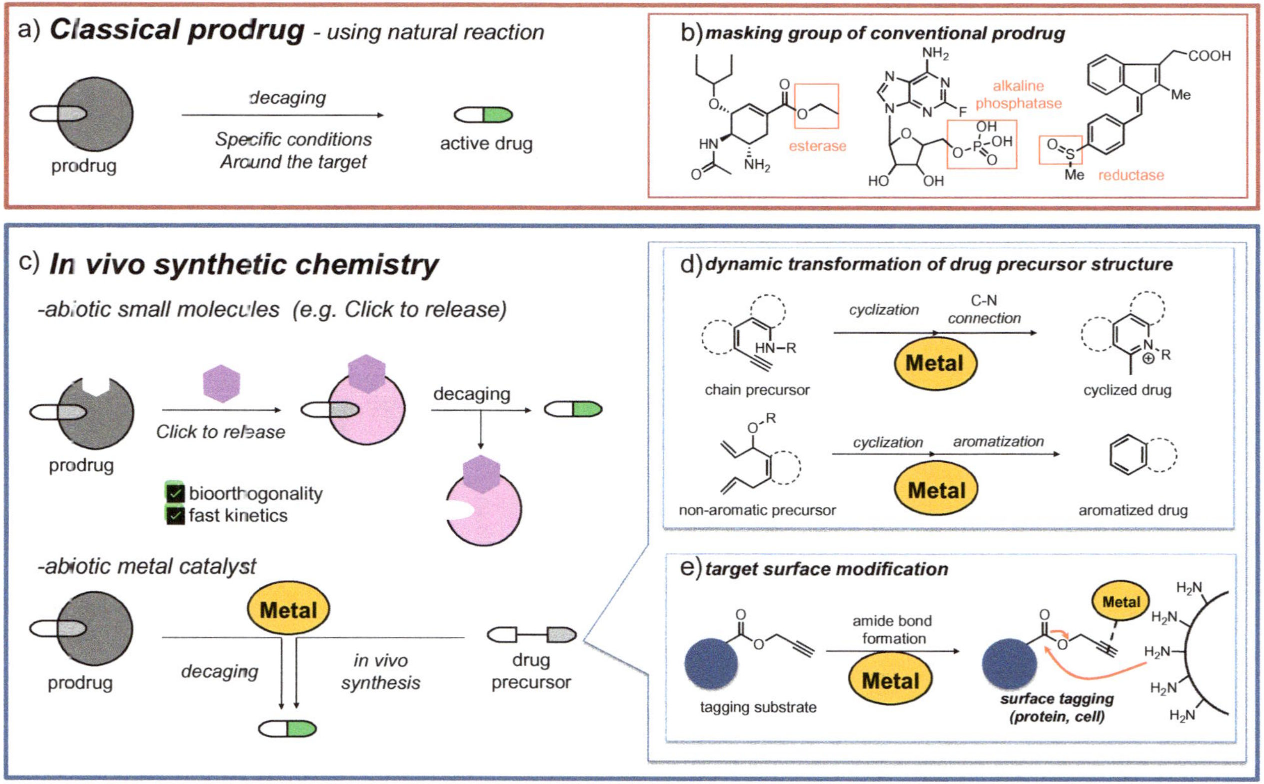

a) Classical prodrug - *using natural reaction*

prodrug → *decaging* → active drug

*Specific conditions
Around the target*

b) masking group of conventional prodrug

esterase

alkaline phosphatase

reductase

c) In vivo synthetic chemistry

-abiotic small molecules (e.g. Click to release)

Click to release

▷ bioorthogonality
▷ fast kinetics

prodrug → *decaging* → active drug

-abiotic metal catalyst

prodrug → *decaging* → active drug

Metal → *in vivo synthesis* → drug precursor

d) dynamic transformation of drug precursor structure

chain precursor

cyclization | Metal | *C–N connection*

cyclized drug

non-aromatic precursor

cyclization | Metal | *aromatization*

aromatized drug

e) target surface modification

tagging substrate

Metal *amide bond formation*

surface tagging (protein, cell)

Fig. 1 (A, B) Classical prodrug activation strategy using natural reactions and (C–E) in vivo synthetic chemistry using abiotic small molecules or abiotic metal catalysts.

To demonstrate therapeutic in vivo synthetic chemistry in the complex environment of living organisms, the reactions need to be designed to occur only at the target site, which is an unprecedented challenge in conventional organic synthetic chemistry. In order to overcome this challenge, we combined two technologies[1]: a targeting method using glycan-modified human serum albumin (glycoHSA), and[2] biocompatible metal-catalyzed reactions with an artificial metalloenzyme (ArM) derived from HSA. Herein this review summarizes our recent progress in therapeutic in vivo synthetic chemistry for cancer therapy using targeted-metal catalyzed reactions.

2. Glycan targeting

Among the various targeting techniques which have been developed to efficiently treat diseases, the reduction of off-target effects in cancer therapy is the most studied.[8] A variety of unique characteristics of various cancer cells, including the morphology of organs[9] and protein expression profiles,[10,11] are utilized for distinguishing them from normal cells. The interaction of peptide/protein or nucleic acid with the target molecule, represented by the aptamer or antibody,[12–14] is crucial for the cancer targeting rather than the weak interaction of glycans.

The interactions which occur through glycans play an important role in the cell recognition process in nature (Fig. 2), and lectin, a protein that binds to glycan, is partially responsible for the interaction network as it exists near the cell surface as a membrane-associated or soluble protein. Since the interaction between a single glycan and lectin is not strong, the glycans exist on the cell surface as an aggregate (including glycoprotein, glycolipid, and proteoglycan, which are collectively called the glycocalyx) to multivalently and heterogeneously enhance the interaction, resulting in strong cell–cell communication.[15] The glycocalyx is utilized in biological systems to distinguish the different types of cells and discriminate viruses and pathogens from the body's own cells, because the glycocalyx acts as the "face" of individual cells, and the glycan pattern changes depending on the type of cells, diseases, and other conditions. We refer to recognition through multivalent and heterogenous interactions with glycans as "glycan pattern recognition."[16] Mimicking the glycan pattern recognition seems to be a promising method to provide high selectivity in the application of this targeting method, but prior to our research, there were few examples of modified proteins with multiple types of glycans for which to use this pattern recognition process. There are some examples that use only a single glycan or sugar to

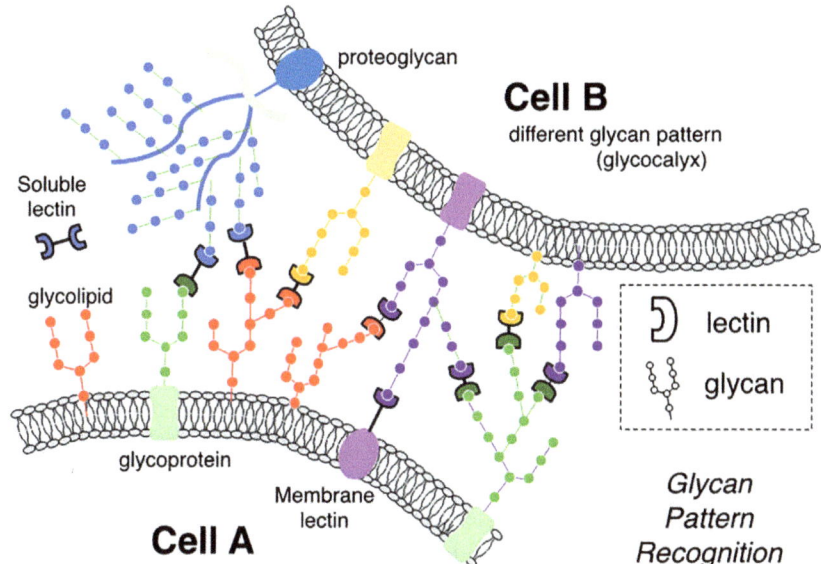

Fig. 2 Cell–cell interaction via a variety of surface glycans and lectins through the glycan pattern-recognition process.

express the target recognition ability, exemplified with the fluorophores or drugs modified with CD22 interactive glycan for delivery to B-lymphoma cells.[17] This difficulty in mimicking the glycan pattern recognition was attributed to the complex structure of glycan assembly with multiple and diverse glycans, which is difficult to prepare with the desired glycan pattern from the viewpoint of organic synthesis.

Previously we have discovered the accelerated 6π-azaelectrocyclization of 1-azatriene by substituent effects (C-4 carbonyl group and C-6 alkenyl or phenyl group), which are called the RIKEN click reaction (Fig. 3).[18] This reaction is carried out smoothly under mild conditions and applies to the protein labeling that selectively conjugates lysine with various functional molecules.[19,20] Modification of huge N-type glycan chains, which is generally thought to be difficult, can also be achieved by the RIKEN click reaction, producing a glycan assembly (i.e., a glycocluster) on molecular surfaces such as dendrimers,[21] lymphocytes,[22] and living cells.[23]

Human serum albumin (HSA), the most abundant protein in the blood, was used as a core unit to produce the glycocluster because HSA has several exposed lysine residues as anchoring sites for glycans. The fluorescence-labeled glycoHSA prepared by the RIKEN click reaction was injected into a normal mouse and monitored to evaluate its biodistribution (Fig. 4A).[24]

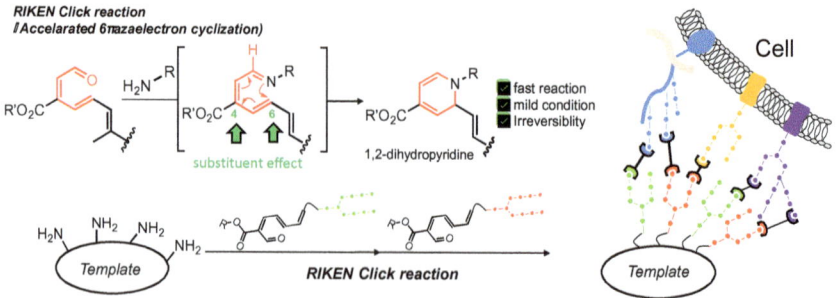

Fig. 3 RIKEN click reaction for glycan modification of the template.

Compared to the fact that HSA without glycan was distributed to the whole body and showed no accumulation, the glycoHSA with glycans **1a–e** demonstrated different behaviors such as organ-specific accumulation and changes in excretion routes. The glycoHSA with glycan **1a** (Man) or **1b** (GlcNAc) selectively accumulated in the mouse liver. Dissection studies revealed that glycoHSA with galactose-terminated glycan **1c** (Gal) was shuttled from the liver toward the gall bladder and intestines. On the other hand, glycoHSA with sialylated glycan **1d** (2,3-Sia) or **1e** (2,6-Sia) was shuttled toward the bladder and kidney. Therefore, the biodistribution of HSA in mice was completely changed depending on the type of glycans attached.

GlycoHSA with two different glycans, prepared by the sequential treatment with RIKEN click reagents possessing different glycan chains, showed different cancer-recognition activity according to the glycan pattern (Fig. 4b).[25] Synthesized HSA with patterns A (**1d** and **1e**), B (**1d** and **1c**), or C (**1d** and **1b**) were evaluated in terms of their recognition ability toward 11 types of cancer cells. In the case of pattern A, glycoHSA specifically recognized HeLa229 cells compared to the other 10 cancer cells. Although glycoHSA with pattern B did not show high cancer recognition of the 11 types of cells, including HeLa229, moderate recognition toward U87MG and DLD1 was shown compared to pattern A. GlycoHSA with pattern C also showed no high cancer recognition properties but did demonstrate moderate recognition properties toward OVCAR3 that the other glycoHSAs did not recognize. Then, applying these different recognition properties among three glycan-modifying patterns, it was shown that glycan pattern recognition could distinguish the type of cancer in vivo using three types of cancer-transplanted mice (HeLa229, U87MG, and DLD1). In the case of pattern A, glycoHSA accumulated in HeLa cancer cells but not in

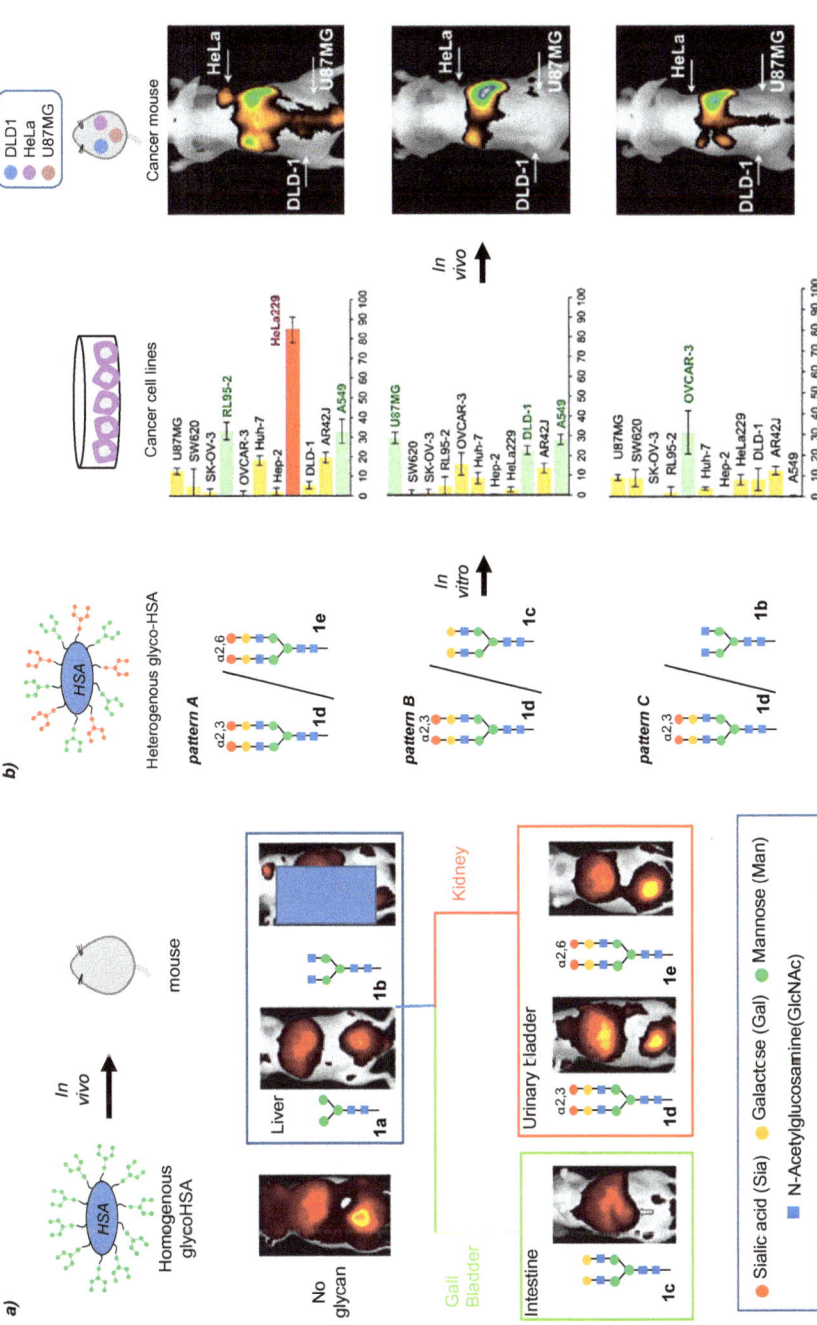

Fig. 4 (A) Biodistribution of homogenous glycoHSA in mice is different depending on the glycan structure. (B) Comparison of interactions between heterogenous glycoHSA with two different types of glycans against 11 cancer lines (in vitro) or three types of cancer-implanted mice (in vivo).

DLD1 or in U87MG cells. The lack of accumulation in the other combinations indicates that the in vivo glycan-dependent accumulation in cancer cells is correlated to the cancer recognition properties evaluated in vitro. This cancer-targeting system using glycan pattern recognition also has advantages compared to current techniques (i.e., antibody targeting) in terms of in vivo stability and faster accumulation on target sites.

These excellent features make this targeting technology promising for the diagnosis and treatment of diseases based on in vivo synthetic chemistry, which is needed to produce the site-specific reactions in vivo. By synthesizing a comprehensive library of glycans and analyzing their interactions with abnormal cells including cancer cells, we will be able to further expand the scope of the application of medicinal technology based on glycan pattern recognition. For this purpose, we have been synthesizing an HSA-modifying probe by pre-linking two or more glycans, enabling more precisely defined glycan patterns.[26] By utilizing these methods to synthesize a library with a wide range of glycan patterns, it will be possible to identify abnormal cells other than cancer cells.

3. In vivo stabilization of a transition metal catalyst as an artificial metalloenzyme (ArM)

One of the promising applications of the glycoHSA should be its application to artificial metalloenzymes (ArMs), which is defined as the incorporation of abiotic transition metals into a protein scaffold.[7,27–29] ArMs are highly chemoselective and act as orthogonal metal catalysts in biological settings. The targeting ability of glycoHSA is a unique advantage as a protein scaffold of ArMs, as it realizes in vivo synthetic chemistry using transition metal catalysts. Additionally, the stabilizing effect of metal catalysts was found in our study to be provided by the hydrophobic pocket of HSA, while the metal catalyst itself was generally unstable under biological conditions due to the existence of abundant quenchers such as glutathione.

HSA is known as a transport protein in the human body, and it has many kinds of binding pockets that accept various ligands.[30] Therefore, the metal complex of HSA can be formed by the conjugate molecule of the metal catalyst and the ligand, and also the glycoHSA. The ArM of HSA was prepared by utilizing diethylaminocoumarin as an anchoring unit for the hydrophobic binding pocket of HSA (drug site I), and complex formation can be confirmed by the change in the fluorescent properties of the coumarin unit

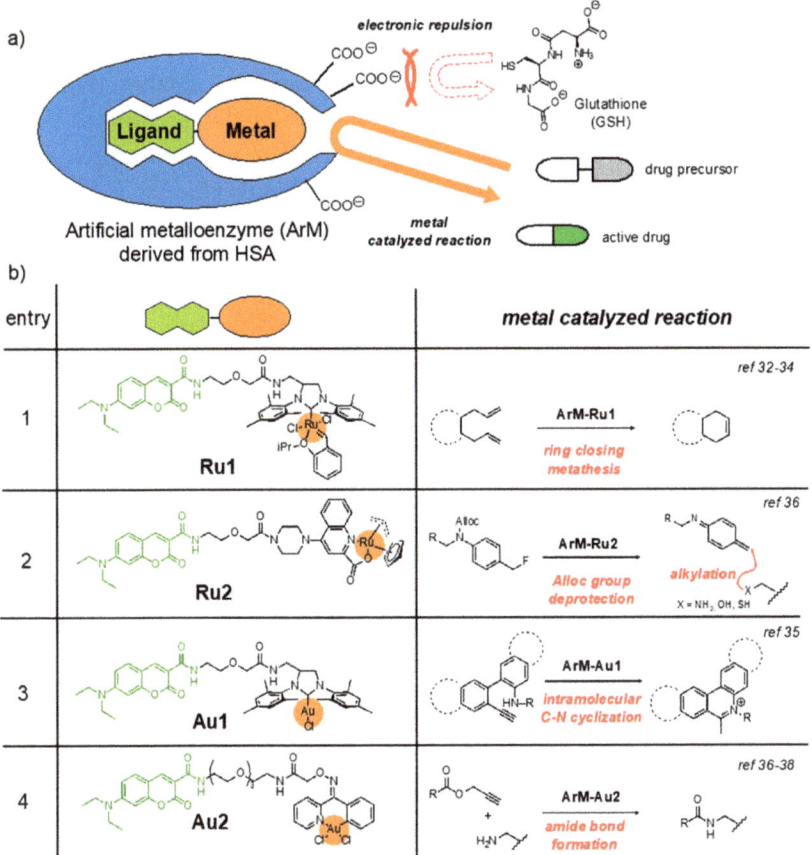

Fig. 5 (A) Protection of the transition metal catalyst inside the HSA pocket from catalyst quenchers such as glutathione (GSH). (B) Whole structures of conjugate metal catalyst and diethylaminocoumarin ligand and reaction catalyzed by the catalyst.

induced by ligand binding.[31] Thus far, we have prepared and evaluated four kinds of ArMs using glycoHSA (or HSA) as a protein scaffold (Fig. 5B).[32-38]

The first example of our ArM is the HSA complex of **Ru1**(**ArM-Ru1**), which catalyzed a metathesis reaction (Fig. 5b, entry 1).[32] In the reaction using the 1,6-heptadiene derivative to produce cyclopentenes, the ArM-Ru1 complex thus prepared showed a turnover number (TON) of approximately 15. Importantly, the catalytic activity was not significantly changed in the presence of 20 mM GSH (a biological quencher, logP −1.30), while the catalyst was easily quenched by the hydrophobic thiol compound, dodecanthiol (an abiotic quencher, logP 7.13). This result can be explained by the negatively charged surface around the site I pocket of HSA at physiological pH, which can ionically repulse GSH from entering the hydrophobic binding pocket to destroy the Ru1 catalyst (Fig. 5A). In addition, further study revealed that the catalytic reaction of ArM–Ru1 was carried out even in blood, which is proof of the in vivo applicability of the metal catalyst through a complex formed with HSA.[33]

The protective effect of the metal catalyst was also exerted in the other ArMs of HSA. In the case of gold-catalyzed hydroamination (Fig. 5B, entry 3), the naked catalytic unit of **Au1** could convert 2′-alkynyl-N-methyl-2-biphenylamine to the 5-methylphenanthridinium derivatives with high TON (around 400), but the catalytic reaction was completely prevented by GSH (1.2 mM).[35] In this case, other biomolecules such as lysine or glucose, did not affect the catalytic activity. On the contrary, the **ArM–Au1** was still active with moderate TON (>100) with GSH treatment, while the catalytic activity without quenchers (TON ~200) was decreased compared to the native catalyst because of decreased accessibility of the substrate to the catalytic center. As another example, **Ru2**, which catalytically detached the allyloxycarbonyl (Alloc) group (Fig. 5B, entry 2), is significantly deactivated by preincubation with even 200 μM of GSH for 30 min (TON with or without GSH treatment = 6.8 or 0.5 h^{-1}, respectively),[36] as demonstrated by several studies performed using the catalyst in a cellular environment.[39–43] On the contrary, **ArM–Ru2** with a surface-modified HSA was still active (TON 11.3 h^{-1}) after incubating with GSH. Interestingly, **ArM–Ru2** has a unique activity in the presence of GSH, likely due to the scavenging effect of the allyl moiety of the π-allyl complex.[43]

Based on these results, it was revealed that glycoHSA possessed not only targeting ability but also the ability to protect the susceptible metal catalyst with the hydrophobic pocket. Therefore, the metal complex of glycoHSA (glycan modified artificial metalloenzyme; GArM) can play a pivotal role in the therapeutic in vivo synthetic chemistry to induce the abiotic reaction in biological systems. While the application of the GArM is not limited to medicinal use—an HSA-based ArM biosensor was actually developed for the visualization of ethylene gas as a plant hormone[34]—this review focuses on tumor therapy using the strategy of in vivo drug synthesis or cell tagging on the basis of in vivo synthetic chemistry (Fig. 6).

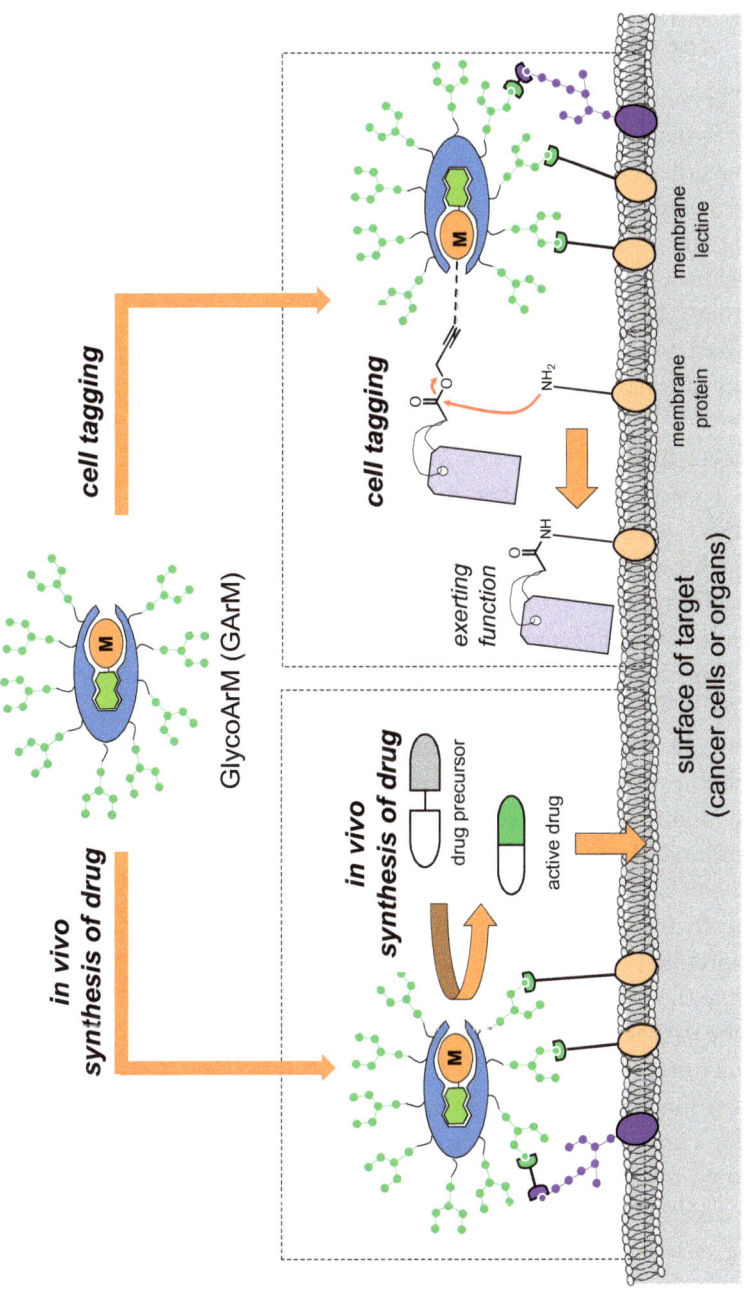

Fig. 6 Tumor therapeutic strategy by in vivo synthesis of drug or cell tagging on the basis of in vivo synthetic chemistry.

4. Antitumor drug synthesis using a metal catalyst for therapeutic in vivo synthetic chemistry

The findings of the superior properties of the glycoHSA scaffold brought us to the next step of therapeutic in vivo synthetic chemistry involving the synthesis of bioactive drugs on cells or in mice for cancer treatment. Abiotic metal catalysts can realize drastic structural transformation of drug precursors to generate active drugs. Therefore, the precursor can be designed from active drugs without generally modifiable functional groups (e.g., amino, hydroxy, or carboxy groups). In our research, the drugs were synthesized from unique precursors, with all of them not being suitable for conventional prodrug strategy. **GArM-Ru1**-catalyzed ring-closing metathesis[32,33] and **ArM-Au1**-catalyzed hydroamination[35] were used to treat cancer through in vitro or in vivo experiments.

To develop a therapeutic method for in vivo drug synthesis using glycoHSA, glycoHSA (2,3-Sia) was applied to deliver the target tumor site, which can recognize HeLa, A549, and SW620 tumors with moderate to high affinity. During the initial research of drug synthesis via ring-closing metathesis, umbelliprenin (**2a**), a natural product extracted from *Ferula* plants, was chosen as a potential antitumor drug that could be activated (Fig. 7A).[32,44] In addition, the hydrophobic feature of the umbelliprenin precursor **3a** was suitable for reaction with HSA-based ArMs in their hydrophobic pockets. The hydrophobic **3a** precursor with **ArM-Ru1** showed significantly higher activity ($k_{cat}/K_M = 13.6\,M^{-1}\,s^{-1}$) compared to compound **3b** without an alkyl chain ($k_{cat}/K_M = 0.009\,M^{-1}\,s^{-1}$), likely because the hydrophobicity of the long alkyl chain of **3a** promotes entry into the hydrophobic-binding pocket of HSA. The antitumor effect of umbelliprenin synthesis was evaluated against three cancer cell lines to determine the cellular activity of GArM-based in vivo synthetic chemistry (Only the data against SW620 is shown in Fig. 7B). For all three cancer cell lines, mixtures of prodrug **3a** and **GArM(2,3-Sia)-Ru1** could induce concentration-dependent significant decreases in cell growth (<5%). Since the combination of prodrug **3a/ArM–Ru1** without glycan vectors was shown to have much fewer cytotoxic effects, this would suggest that glycan targeting plays a key role in the cytotoxic effects of GArM-based prodrug activation, due to surface-localized or intracellular prodrug activation.

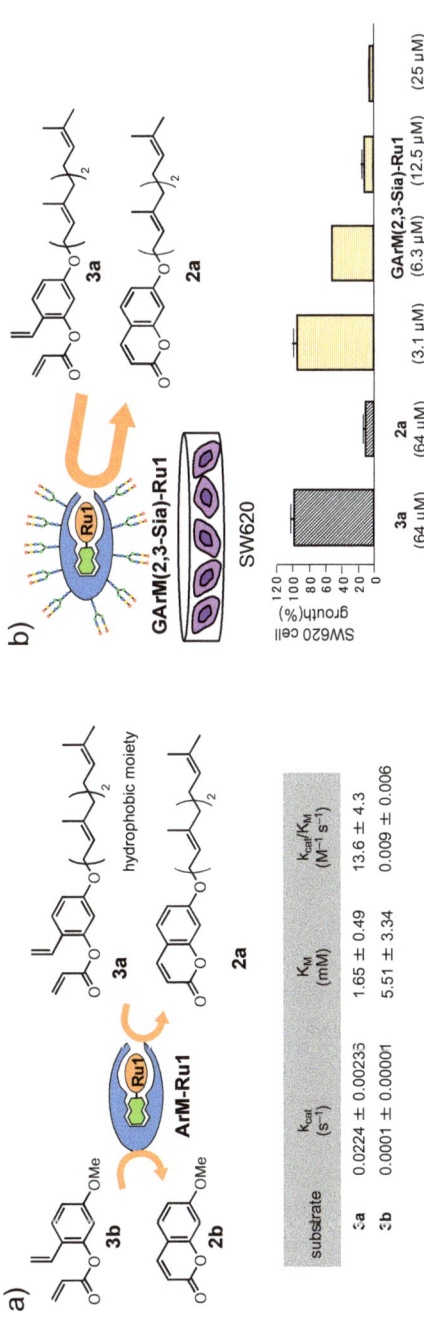

Fig. 7 (A) Ring-closing metathesis of **3a** and **3b** catalyzed by **ArM-Ru1**. (B) Synthesis of umbelliprenin by **GArM-Ru1** in three cancer cell lines induced significant decreases in cell growth.

The **GArM(2,3–Sia)–Ru1** can be applied to benzene ring synthesis by sequential ring-closing metathesis and aromatization (Fig. 8).[33,45–47] Phenyl rings were included in many pharmacologically active compounds, so this strategy is widely applicable in drug synthesis.[48] Then, we selected the naphthyl derivative of Combretastatin A-4 (**5**)[49,50] as an antitumor drug and the corresponding designed drug precursor **4a–c**. The leaving group (R) affected the catalytic activity of **ArM–Ru1** because the accessibility of the substrate to the catalytic center and the hydrophobic pocket of HSA was changed. The pivaloyl group in substrate **4a** provided better hydrolysis resistance (hydrolysis of 9.8% in 50% blood for 2h) and k_{cat}/K_M parameters ($457.9\,M^{-1}\,s^{-1}$). These findings can produce efficient drug synthesis in vivo and result in the reduction of catalyst loading which potentially reduces adverse effects. The catalytic system for drug synthesis was applied to in vivo cancer treatment by sequential intravenous injection of substrate **4a** and **GArM(2,3–Sia)–Ru1** to the HeLa cell-bearing mouse. While monitoring tumor growth over a period of 20 days, a clear depreciation in the rate of tumor growth can be seen for the treatment group compared to the controls. These results are one of the successful outcomes of therapeutic in vivo synthetic chemistry for cancer treatment.

An alternative drug synthesis is **Au1**-catalyzed hydroamination of 2′-alkynyl-N-methyl-2-biphenylamine (**6**) toward 5-methylphenanthridinium derivatives **7** (Fig. 9),[35] of which the scaffold was included in several antitumor DNA-intercalators.[51] The cytotoxicity of precursor **6** against A549 cells was evaluated under the conditions with **Au1** or **ArM–Au1**. Although concentration–dependent toxicity against A549 cells was observed in both cases using **Au1** or **ArM–Au1** (Fig. 9), the combination of precursor **6** and **ArM–Au1** showed more potent toxicity. The usage of $0.63\,\mu M$ **ArM–Au1** decreased cell growth by 50%, whereas $2.5\,\mu M$ **Au-1** was required to achieve the same level of growth inhibition. In addition, **ArM–Au1** was less toxic at $10\,\mu M$ than **Au1** which exerted obvious cytotoxicity at the same concentration, implying that the ArM could avoid unwanted side effects. These observations highlight the potential importance of HSA-based ArMs as triggers for therapeutic in vivo synthetic strategies. Although HSA was utilized as is for the protein scaffold in the study regarding the **Au1** catalyst, these results suggested the potential success of the in vivo application of the direct synthesis of a **7** by an **Au1**-catalyzed reaction in combination with glycoHSA.

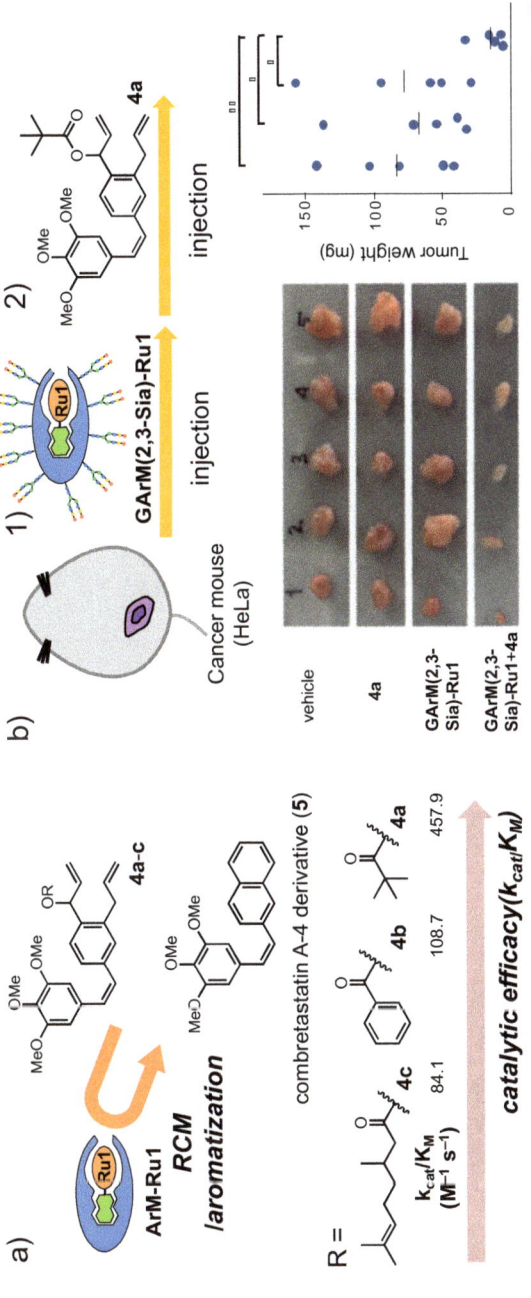

Fig. 8 (A) Naphthyl derivative of Combretastatin A-4 (**5**) synthesis from non-aromatic precursor **4a−c** catalyzed by **ArM-Ru1** and the influence of the ester moiety toward affinity of **4a−c** to the hydrophobic pocket of HSA and reaction kinetics. (B) In vivo synthesis of drug **5** and sequential injection of **GArM(2,3-Sia)-Ru1** and substrate **4a** significantly decreased tumor volumes and tumor weight. *$P < 0.05$, **$P < 0.01$.

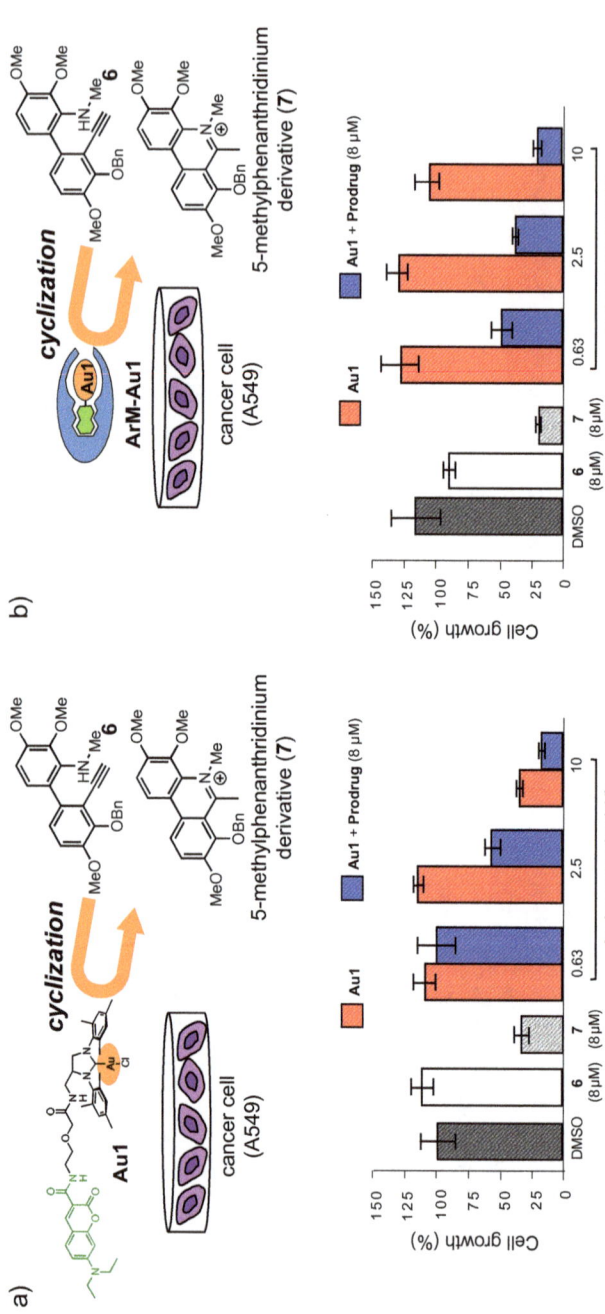

Fig. 9 Cytotoxicity assay with the synthesis of 5-methylphenanthridinium derivative **7** from precursor **6** catalyzed by (A) **Au1** or (B) **ArM-Au1**.

5. Selective cell tagging (SeCT) therapy

The second strategy of therapeutic in vivo synthetic chemistry using the GArMs is the application of the catalytic covalent bond formation reaction of biofunctional molecules to the target cell/organ to treat the disease. This is called selective cell tagging (SeCT) therapy (Fig. 6).[36–38] Cancer treatment would be achieved by exerting the biological function of the covalently tagged molecule, as covalent drugs show stronger activity compared to drugs with a non-covalent interaction.[52] This SeCT concept does not utilize genetic[53] or metabolic pathways,[54] and should therefore apply to medicinal use to exert a therapeutic effect without affecting surrounding tissues by impairing cellular function.

For metal-catalyzed tagging chemistry, we developed a gold-mediated amide bond formation reaction, which is proceeded even under biological conditions by the combination of **ArM-Au2** (Fig. 5, entry 4) and a propargyl ester (PE) substrate.[37,55] We first achieved in vivo site-specific catalytic tagging of fluorophores with the injection of **GArM-Au2** and PE substrate into the normal mouse (Fig. 10). The in vivo fluorescence distribution was changed depending on the attached glycan of glycoHSA. In the case of **GArM(2,6–Sia)–Au2**, which accumulates in the liver, the PE substrate was activated by **GArM(2,6–Sia)–Au2** to react with lysine residues on the surface of liver cells. On the other hand, **GArM(Gal)–Au2** accumulated at the intestine-induced fluorescent labeling of intestine cells. This targeted fluorescent labeling reaction is the first example of an organ-selective tagging reaction in animals.

Encouraged by this successful in vivo tagging reaction, the reaction was next applied to tumor therapy based on an SeCT strategy using a cyclic RGD (cRGD) peptide as a tagged molecule that can disrupt cancer cell adhesion.[38] The cRGD peptide can block the integrins overexpressed on cancer cells, leading to tumor regression through inhibition of angiogenesis and prevention of their attachment to the extracellular matrix.[56,57] The tagging reaction of **cRGD-PE** with **GArM(2,3–Sia)–Au2** was expected to inhibit tumor onset without toxicity, since the reaction affects cell adhesion (Fig. 11C), but not cell growth (Fig. 11B). In the in vivo experiment, mice treated with sequential injections of **GArM(2,3–Sia)–Au2** and cRGD-PE showed a better survival rate of 40% at 81 days when injected into HeLa-Luc cells, while the survival rates decreased to 0% for other control conditions before the final time point. In this manner, the effect of integrin-blocking

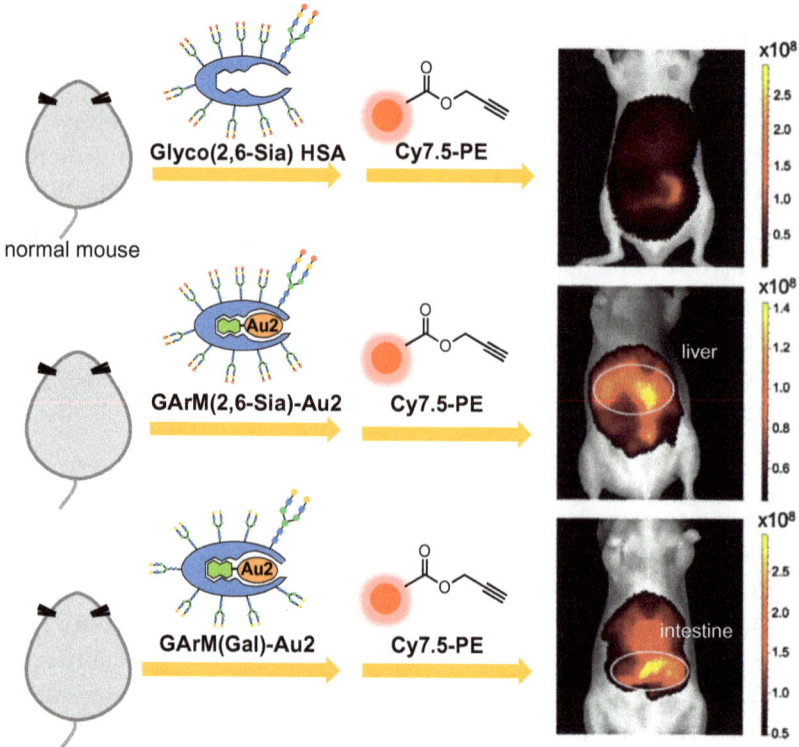

Fig. 10 In vivo site-specific catalytic tagging of fluorophores using a gold-mediated amide bond formation reaction in living mice.

SeCT therapy can be monitored well before cancer cells are given time to seed onto the extracellular matrix. Overall, these data suggest that tumor onset and progression in mice receiving cRGD-based SeCT therapy had been significantly impaired.

The ideal substrate in the SeCT strategy only exerts cytotoxicity when it is covalently attached to the cellular surface but is otherwise nontoxic. We serendipitously found the proapoptotic peptide **8** (Ac-GGKLFG-PE), which exerted cytotoxicity against SW620 cells only in the presence of the **Au2** complex of cRGD-coated HSA [HSA(cRGD)], which is also a promising protein scaffold for in vivo targeting in addition to GlycoHSA (Fig. 12A).[36] The combination of peptide **8** and **ArM(cRGD)–Au2** results in good tumor growth inhibition in both in vitro and in vivo assays, indicating that the observed cytotoxicity is based on the synergistic effects of the tagging reaction.

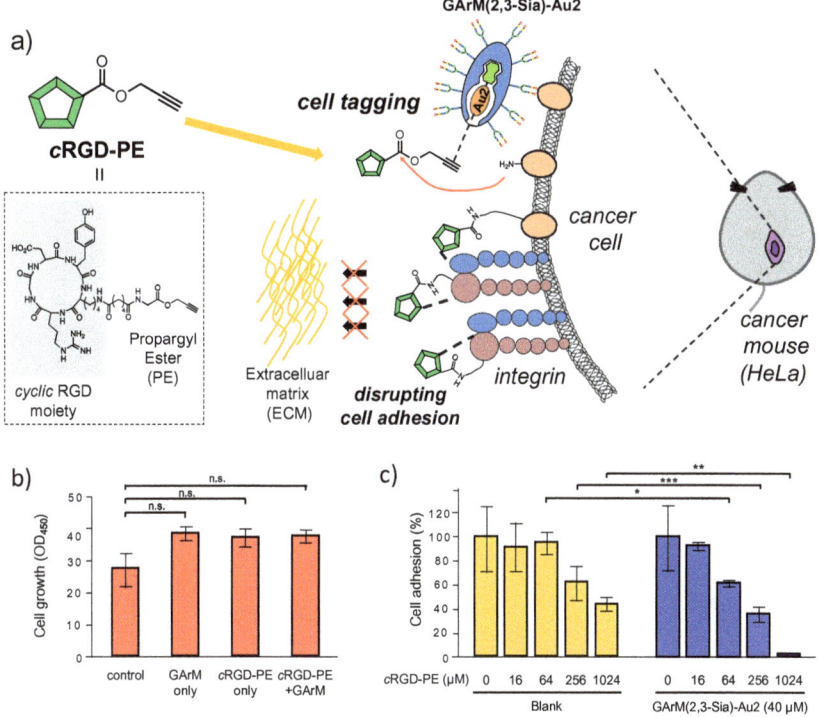

Fig. 11 (A) Cell tagging of *c*RGD peptide using **cRGD-PE** to the cancer cell surface, disrupting cell adhesion. (B and C) Cell tagging of *c*RGD peptide on the cancer cell surface induced cell-adhesion inhibition rather than cell-growth inhibition. n.s. not significant, $*P < 0.05$, $**P < 0.01$, $***P < 0.005$.

Primary SeCT therapy has been studied by using only the combination of the **Au2** complex and propargyl ester substrate, but we found that **Ru2**-catalyzed alkylation using *p*-Alloc-protected aminobenzyl fluoride (BnF) substrate **9** as a tagging reaction showed better efficiency (Fig. 5, entry 2; Fig. 12).[36] The proapoptotic peptide substrate **9** worked in the CpRu/BnF tagging system, while the modified form should be different from the **Au2**-catalyzed amidation. Cancer therapy was conducted by administering **ArM(cRGD)–Ru2** and peptide **9** to SW620 tumor-bearing mice (Fig. 12b). The compounds were intravenously injected only on the first day, in contrast to the experiment using **ArM(cRGD)–Au2** and peptide **8**, which required frequent injections for 10 consecutive days. As the result, co-injection of **ArM(cRGD)–Ru2** and peptide **9** significantly and synergistically suppressed tumor growth, differently from peptide **9** or catalyst alone, which did not inhibit tumor growth. Therefore, cancer SeCT

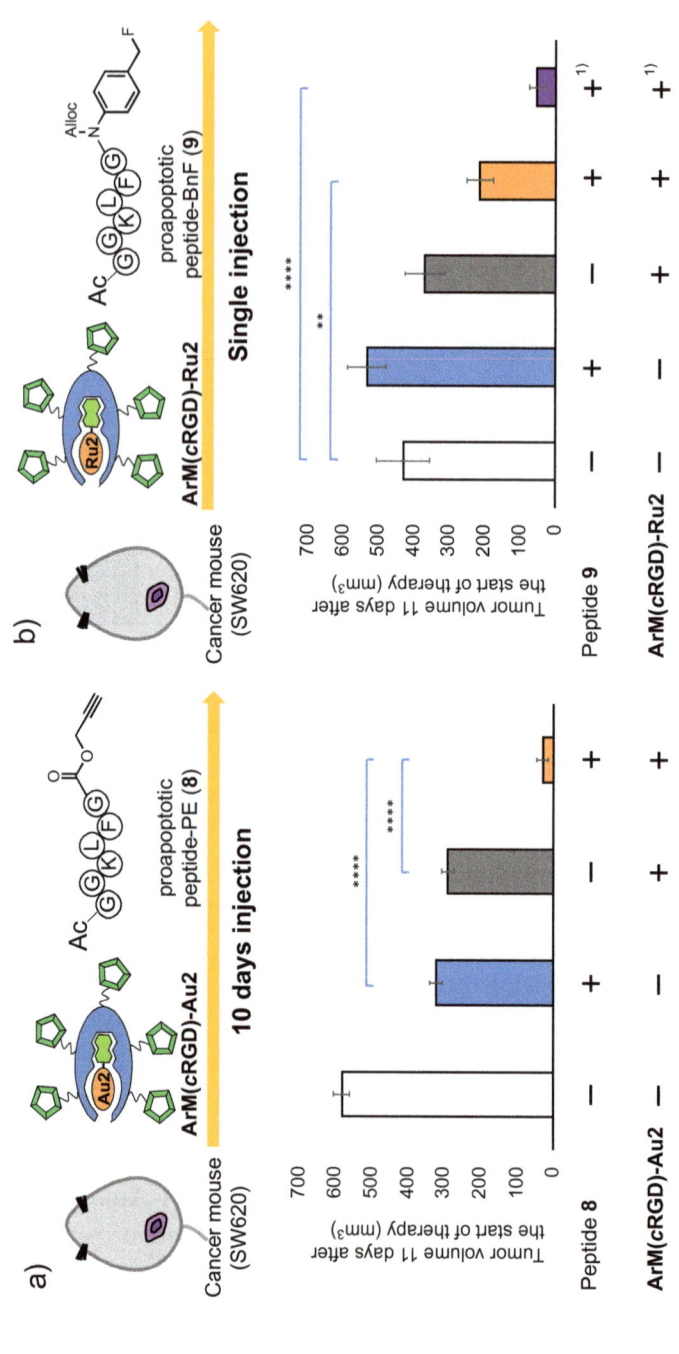

Fig. 12 (A) In vivo cell tagging of proapoptotic peptide **8** to the cancer cell by **ArM(cRGD)-Au2**. (B) In vivo cell tagging of proapoptotic peptide **9** to the cancer cell by **ArM(cRGD)-Ru2**. (1) twice the amount of compounds were injected to mice. **$P < 0.01$, ****$P < 0.001$.

therapy was achieved using the **Ru2** catalytic system with superior therapeutic results and only a single injection of compounds via a tail vein. This is attributed to the higher efficacy of the tagging reaction and lower toxicity of the **ArM(cRGD)-Ru2** catalyst.

6. Conclusions

For establishing the concept of therapeutic in vivo synthetic chemistry, we have studied two aspects of glycoHSA: the cancer-targeting ability of glycan pattern recognition and the stabilization of transition-metal catalysts inside the hydrophobic pocket. The combination of these two properties of glycoHSA enabled GArMs to deliver metal catalysts to the target site without deactivation of the metal catalyst for in vivo synthetic chemistry. We demonstrated two strategies of therapeutic in vivo synthetic chemistry: (1) in vivo drug synthesis by drastic structural transformation of the drug precursor, which is different from conventional prodrug strategies that rely on de-caging of the masking group; and (2) SeCT strategy by tagging the cancer surface of functional molecules, which allows the drug to exert its function only on the tagged cell. The effectiveness of in vivo synthetic chemistry using these two strategies was demonstrated in cancer therapy through transition metal-catalyzed reactions in cellular systems and in mice. In the future, the increasing complexity of glycan patterns of glycoArM will allow for more selective catalyst delivery to various types of organs and cells, and more diverse catalysts can be used to achieve drastic and flexible transformations. These studies of the concept of therapeutic in vivo synthetic chemistry will lead to not only the treatment of diseases other than cancer, but also to the development of tailor-made therapies.

Acknowledgments

We would like to thank Glytech, Inc. for supplying various N-glycans. This work was financially supported by the AMED Grant JP15KM0908001, a research grant from the Astellas Foundation, Mizutani Foundation for Glycoscience, and JSPS KAKENHI Grant Numbers, JP21H02065, JP21K19042 (to K.T.), JP20K15968 and JP19J00396(to K.M.).

References

1 Rautio, J.; Kumpulainen, H.; Heimbach, T.; Oliyai, R.; Oh, D.; Järvinen, T.; Savolainen, J. Prodrugs: Design and Clinical Applications. *Nat. Rev. Drug Discov.* **2008**, 7, 255–270.
2. Rautio, J.; Meanwell, N. A.; Di, L.; Hageman, M. J. The Expanding Role of Prodrugs in Contemporary Drug Design and Development. *Nat. Rev. Drug Discov.* **2018**, 17, 559–587.

3. Versteegen, R. M.; Rossin, R.; ten Hoeve, W.; Janssen, H. M.; Robillard, M. S. Click to Release: Instantaneous Doxorubicin Elimination upon Tetrazine Ligation. *Angew. Chem., Int. Ed.* **2013**, *52*, 14112–14116.

4. Ji, X.; Pan, Z.; Yu, B.; De La Cruz, L. K.; Zheng, Y.; Ke, B.; Wang, B. Click and Release: Bioorthogonal Approaches to "On Demand" Activation of Prodrugs. *Chem. Soc. Rev.* **2019**, *48*, 1077–1094.

5. Pradipta, A. R.; Ahmadi, P.; Terashima, K.; Muguruma, K.; Fujii, M.; Ichino, T.; Maeda, S.; Tanaka, K. Targeted 1,3-Dipolar Cycloaddition with Acrolein for Cancer Prodrug Activation. *Chem. Sci.* **2021**, *12*, 5438–5449.

6. Chang, T.-C.; Tanaka, K. In Vivo Organic Synthesis by Metal Catalysts. *Bioorg. Med. Chem.* **2021**, *46*, 116353.

7. Tanaka, K.; Vong, K. The Journey to In Vivo Synthetic Chemistry: Azaelectrocyclization to Artificial Metalloenzymes. *Bull. Chem. Soc. Jpn.* **2020**, *93*, 1275–1286.

8. Rosenblum, D.; Joshi, N.; Tao, W.; Karp, J. M.; Peer, D. Progress and Challenges Toward Delivery of Cancer Chemotherapeutics. *Nat. Commun.* **2018**, *9*, 1410.

9. Shi, Y.; Van der Meel, R.; Chen, X. Lammers, the EPR Effect and beyond: Strategies to Improve Tumor Targeting and Cancer Nanomedicine Treatment Efficacy. *Theranostics* **2020**, *10*, 7921–7924.

10. Wang, J.; Xu, B. Targeted Therapeutic Options and Future Perspectives for HER2-Positive Breast Cancer. *Signal Transduct. Target. Ther.* **2019**, *4*, 34.

11. Denis, M. G.; Bennouna, J. Osimertinib for Front-Line Treatment of Locally Advanced or Metastatic *EGFR*-Mutant NSCLC Patients: Efficacy, Acquired Resistance and Perspectives for Subsequent Treatments. *Cancer Manag. Res.* **2020**, *12*, 12593–12602.

12. Alshaer, W.; Hillaireau, H.; Fattal, E. Aptamer-Guided Nanomedicines for Anticancer Drug Delivery. *Adv. Drug Deliv. Rev.* **2018**, *134*, 122–137.

13. Khongorzul, P.; Ling, C. J.; Khan, F. U.; Ihsan, A. U.; Zhang, J. Antibody–Drug Conjugates: A Comprehensive Review. *Mol. Cancer Res.* **2020**, *18*, 3–19.

14. Awwad, S.; Angkawinitwong, U. Overview of Antibody Drug Delivery. *Pharmaceutics* **2018**, *10*, 83.

15. Tarbell, J. M.; Cancel, L. M. The Glycocalyx and its Significance in Human Medicine. *J. Intern. Med.* **2016**, *280*, 97–113.

16. Kurbangalieva, A.; Zamalieva, R.; Nasibullin, I.; Yamada, K.; Tanaka, K. Homo- and Heterogeneous Glycoconjugates on the Basis of N-Glycans and Human Serum Albumin: Synthesis and Biological Evaluation. *Molecules* **2022**, *27*, 1285.

17. Peng, W.; Paulson, J. C. CD22 Ligands on a Natural N-Glycan Scaffold Efficiently Deliver Toxins to B-Lymphoma Cells. *J. Am. Chem. Soc.* **2017**, *139*, 12450–12458.

18. Tanaka, K.; Mori, H.; Yamamoto, M.; Katsumura, S. Significant Acceleration of 6π-Azaelectrocyclization Resulting from a Remarkable Substituent Effect and Formal Synthesis of the Ocular Age Pigment A2-E by a New Method for Substituted Pyridine Synthesis. *J. Organomet. Chem.* **2001**, *66*, 3099–3110.

19. Tanaka, K.; Masuyama, T.; Hasegawa, K.; Tahara, T.; Mizuma, H.; Wada, Y.; Watanabe, Y.; Fukase, K. Submicrogram-Scale Protocol for Biomolecule-Based PET Imaging by Rapid 6π-Azaelectrocyclization: Visualization of Sialic Acid Dependent Circulatory Residence of Glycoproteins. *Angew. Chem., Int. Ed.* **2008**, *47*, 102–105.

20. Ogura, A.; Tahara, T.; Nozaki, S.; Morimoto, K.; Kizuka, Y.; Kitazume, S.; Hara, M.; Kojima, S.; Onoe, H.; Kurbangalieva, A.; Taniguchi, N.; Watanabe, Y.; Tanaka, K. Visualizing Trimming Dependence of Biodistribution and Kinetics with Homo- and Heterogeneous N-Glycoclusters on Fluorescent Albumin. *Sci. Rep.* **2016**, *6*, 21797.

21. Tanaka, K.; Siwu, E. R. O.; Minami, K.; Hasegawa, K.; Nozaki, S.; Kanayama, Y.; Koyama, K.; Chen, W. C.; Paulson, J. C.; Watanabe, Y.; Fukase, K. Noninvasive Imaging of Dendrimer-Type N-Glycan Clusters: In Vivo Dynamics Dependence on Oligosaccharide Structure. *Angew. Chem., Int. Ed.* **2010**, *49*, 8195–8200.

22. Tanaka, K.; Minami, K.; Tahara, T.; Siwu, E. R. O.; Koyama, K.; Nozaki, S.; Onoe, H.; Watanabe, Y.; Fukase, K. A Combined 6π-Azaelectrocyclization/Staudinger Approach to Protein and Cell Engineering: Noninvasive Tumor Targeting by N-Glycan-Engineered Lymphocytes. *J. Carbohydr. Chem.* **2010**, *29*, 118–132.

23. Tanaka, K.; Kitadani, M.; Tsutsui, A.; Pradipta, A. R.; Imamaki, R.; Kitazume, S.; Taniguchi, N.; Fukase, K. A Cascading Reaction Sequence Involving Ligand-Directed Azaelectrocyclization and Autooxidation-Induced Fluorescence Recovery Enables Visualization of Target Proteins on the Surfaces of Live Cells. *Org. Biomol. Chem.* **2014**, *12*, 1412–1418.

24. Ogura, A.; Tahara, T.; Nozaki, S.; Onoe, H.; Kurbangalieva, A.; Watanabe, Y.; Tanaka, K. Glycan Multivalency Effects toward Albumin Enable N-Glycan-Dependent Tumor Targeting. *Bioorg. Med. Chem. Lett.* **2016**, *26*, 2251–2254.

25. Ogura, A.; Urano, S.; Tahara, T.; Nozaki, S.; Sibgatullina, R.; Vong, K.; Suzuki, T.; Dohmae, N.; Kurbangalieva, A.; Watanabe, Y.; Tanaka, K. Visible Strategy for Screening the Effects of Glycan Heterogeneity on Target Organ Adhesion and Biodistribution in Live Mice. *Chem. Commun. (Cambridge)* **2018**, *54*, 8396–8693.

26. Smirnov, I.; Sibgatullina, R.; Urano, S.; Tahara, T.; Ahmadi, P.; Watanabe, Y.; Pradipta, A. R.; Kurbangalieva, A.; Tanaka, K. A Strategy for Tumor Targeting by Higher-Order Glycan Pattern Recognition: Synthesis and In Vitro and In Vivo Properties of Glycoalbumins Conjugated with Four Different N-Glycan Molecules. *Small* **2020**, *16*, 2004831.

27. Jeschek, M.; Panke, S.; Ward, T. R. Artificial Metalloenzymes on the Verge of New-to-Nature Metabolism. *Trends Biotechnol.* **2018**, *36*, 60–72.

28. Schwizer, F.; Okamoto, Y.; Heinisch, T.; Gu, Y.; Pellizzoni, M. M.; Lebrun, V.; Reuter, R.; Köhler, V.; Lewis, J. C.; Ward, T. R. Artificial Metalloenzymes: Reaction Scope and Optimization Strategies. *Chem. Rev.* **2018**, *118*, 142–231.

29. Vong, K.; Nasibullin, I.; Tanaka, K. Exploring and Adapting the Molecular Selectivity of Artificial Metalloenzymes. *Bull. Chem. Soc. Jpn.* **2021**, *94*, 382–396.

30. Ghuman, J.; Zunszain, P. A.; Petitpas, I.; Bhattacharya, A. A.; Otagiri, M.; Curry, S. Structural Basis of the Drug-Binding Specificity of Human Serum Albumin. *J. Mol. Biol.* **2005**, *353*, 38–52.

31. Jones, G., II; Jackson, W. R.; Choi, C. Y.; Bergmark, W. R. Solvent Effects on Emission Yield and Lifetime for Coumarin Laser Dyes. Requirements for a Rotatory Decay Mechanism. *J. Phys. Chem.* **1985**, *89*, 294–300.

32. Eda, S.; Nasibullin, I.; Vong, K.; Kudo, N.; Yoshida, M.; Kurbangalieva, A.; Tanaka, K. Biocompatibility and Therapeutic Potential of Glycosylated Albumin Artificial Metalloenzymes. *Nat. Catal.* **2019**, *2*, 780–792.

33. Nasibullin, I.; Smirnov, I.; Ahmadi, P.; Vong, K.; Kurbangalieva, A.; Tanaka, K. Synthetic Prodrug Design Enables Biocatalytic Activation in Mice to Elicit Tumor Growth Suppression. *Nat. Commun.* **2022**, *13*, 39.

34. Vong, K.; Eda, S.; Kadota, Y.; Nasibullin, I.; Wakatake, T.; Yokoshima, S.; Shirasu, K.; Tanaka, K. An Artificial Metalloenzyme Biosensor Can Detect Ethylene gas in Fruits and Arabidopsis Leaves. *Nat. Commun.* **2019**, *10*, 5746.

35. Chang, T.-C.; Vong, K.; Yamamoto, T.; Tanaka, K. Prodrug Activation by Gold Artificial Metalloenzyme-Catalyzed Synthesis of Phenanthridinium Derivatives Via Hydroamination. *Angew. Chem., Int. Ed.* **2021**, *60*, 12446–12454.

36. Ahmadi, P.; Muguruma, K.; Chang, T.-C.; Tamura, S.; Tsubokura, K.; Egawa, Y.; Suzuki, T.; Dohmae, N.; Nakao, Y.; Tanaka, K. In Vivo Metal-Catalyzed SeCT Therapy by a Proapoptotic Peptide. *Chem. Sci.* **2021**, *12*, 12266–12273.

37. Tsubokura, K.; Vong, K. K. H.; Pradipta, A. R.; Ogura, A.; Urano, S.; Tahara, T.; Nozaki, S.; Onoe, H.; Nakao, Y.; Sibgatullina, R.; Kurbangalieva, A.; Watanabe, Y.; Tanaka, K. In Vivo Gold Complex within Live Mice. *Angew. Chem., Int. Ed.* **2017**, *56*, 3579–3584.

38. Vong, K.; Tahara, T.; Urano, S.; Nasibullin, I.; Tsubokura, K.; Nakao, Y.; Kurbangalieva, A.; Onoe, H.; Watanabe, Y.; Tanaka, K. Disrupting Tumor Onset and Growth Via Selective Cell Tagging (SeCT) Therapy. *Sci. Adv.* **2021**, *7*, eabg4038.

39. Hsu, H.-T.; Trantow, B. M.; Waymouth, R. M.; Wender, P. A. Bioorthogonal Catalysis: A General Method to Evaluate Metal-Catalyzed Reactions in Real Time in Living Systems Using a Cellular Luciferase Reporter System. *Bioconjugate Chem.* **2016**, *27*, 376–382.

40. Tomás-Gamasa, M.; Martínez-Calvo, M.; Couceiro, J. R.; Mascareñas, J. L. Transition Metal Catalysis in the Mitochondria of Living Cells. *Nat. Commun.* **2016**, *7*, 12538.

41. Völker, T.; Meggers, E. Chemical Activation in Blood Serum and Human Cell Culture: Improved Ruthenium Complex for Catalytic Uncaging of Alloc-Protected Amines. *Chembiochem* **2017**, *18*, 1083–1086.

42. Miguel-Ávila, J.; Tomás-Gamasa, M.; Mascareñas, J. L. Intracellular Ruthenium-Promoted (2+2+2) Cycloadditions. *Angew. Chem., Int. Ed.* **2020**, *59*, 17628–17633.

43. Völker, T.; Dempwolff, F.; Graumann, P. L.; Meggers, E. Progress Towards Bioorthogonal Catalysis with Organometallic Compounds. *Angew. Chem., Int. Ed.* **2014**, *53*, 10536–10540.

44. Shakeri, A.; Iranshahy, M.; Iranshahi, M. Biological Properties and Molecular Targets of Unbelliprenin—A Mini-Review. *J. Asian Nat. Prod. Res.* **2014**, *16*, 884–889.

45. Sabatino, V.; Rebelein, J. G.; Ward, T. R. "Close-to-Release": Spontaneous Bioorthogonal Uncaging Resulting from Ring-Closing Metathesis. *J. Am. Chem. Soc.* **2019**, *141*, 17048–17052.

46. Samanta, A.; Sabatino, V.; Ward, T. R.; Walther, A. Functional and Morphological Adaptation in DNA Protocells Via Signal-Processing Prompted by Artificial Metalloenzymes. *Nat. Nanotechnol.* **2020**, *15*, 914–921.

47. Yoshida, K.; Horiuchi, S.; Iwadate, N.; Kawagoe, F.; Imamoto, T. An Efficient Route to Benzene and Phenol Derivatives Via Ring-Closing Olefin Metathesis. *Synlett* **2007**, 1561–1562.

48. Roughley, S. D.; Jordan, A. M. The Medicinal Chemist's Toolbox: An Analysis of Reactions Used in the Pursuit of Drug Candidates. *J. Med. Chem.* **2011**, *54*, 3451–3479.

49. Maya, A. B. S.; del Rey, B.; de Clairac, R. P. L.; Caballero, E.; Barasoain, I.; Andreu, J. M.; Medarde, M. Design, Synthesis and Cytotoxic Activities of Naphthyl Analogues of Combretastatin A-4. *Bioorg. Med. Chem. Lett.* **2000**, *10*, 2549–2551.

50. Maya, A. B. S.; Pérez-Melero, C.; Mateo, C.; Alonso, D.; Fernández, J. L.; Gajate, C.; Mollinedo, F.; Peláez, R.; Caballero, E.; Medarde, M. Further Naphthylcombretastatins. An Investigation on the Role of the Naphthalene Moiety. *J. Med. Chem.* **2005**, *48*, 556–568.

51. Tumir, L. M.; Stojkovic, M. R.; Piantanida, I. Come-Back of Phenanthridine and Phenanthridinium Derivatives in the 21st Century. *Beilstein J. Org. Chem.* **2014**, *10*, 2930–2954.

52. Ghosh, A. K.; Samanta, I.; Mondal, A.; Liu, W. R. Covalent Inhibition in Drug Discovery. *ChemMedChem* **2019**, *14*, 889–906.

53. Ernst, R. J.; Krogager, T. P.; Maywood, E. S.; Zanchi, R.; Beránek, V.; Elliott, T. S.; Barry, N. P.; Hastings, M. H.; Chin, J. W. Genetic Code Expansion in the Mouse Brain. *Nat. Chem. Biol.* **2016**, *12*, 776–778.

54. Prescher, J. A.; Dube, D. H.; Bertozzi, C. R. Chemical Remodelling of Cell Surfaces in Living Animals. *Nature* **2004**, *430*, 873–877.

55. Lin, Y.; Vong, K.; Matsuoka, K.; Tanaka, K. 2-Benzoylpyridine Ligand Complexation with Gold Critical for Propargyl Ester-Based Protein Labeling. *Chem. Eur. J.* **2018**, *24*, 10595–10600.

56. Ruoslahti, E. RGD and Other Recognition Sequences for Integrins. *Annu. Rev. Cell Dev. Biol.* **1996**, *12*, 697–715.

57. Desgrosellier, J. S.; Cheresh, D. A. Integrins in Cancer: Biological Implications and Therapeutic Opportunities. *Nat. Rev. Cancer* **2010**, *10*, 9–22.

CHAPTER THREE

Pseudo-glycoconjugates with a C-glycoside linkage

Go Hirai*

Graduate School of Pharmaceutical Sciences, Kyushu University, Fukuoka City, Japan
*Corresponding author: e-mail address: gohirai@phar.kyushu-u.ac.jp

Dedicated to the memory of Prof. Hidetoshi Yamada.

Contents

Abbreviations

B	boat conformation
Bn	benzyl
BOM	benzyloxymethyl-, $PhCH_2OCH_2-$
BT	benzothiazole
Bz	benzoyl-
C	chair conformation
CD1d	a glycoprotein presented on human antigen-presenting cells
DFT	density functional theory (calculation)
FBS	fetal bovine serum
GM3	(*N*-acetylenuraminic acid)-Gal-Glc-ceramide
HF	Hartree–Fock (calculation)
LiHMDS	lithium hexamethyldisilazide; lithium bis(trimethylsilyl)amide
MD	molecular dynamics (calculation)
MM	molecular mechanics (calculation)
MP	4-methoxyphenyl-
OTf	triflate, trifluoromethanesulfonate-, $CF_3S(O)_2O-$
Ph	phenyl-

Advances in Carbohydrate Chemistry and Biochemistry, Volume 82
ISSN 0065-2318
https://doi.org/10.1016/bs.accb.2022.10.002
35

PM3	parametric method 3 (calculation)
TES	triethylsilyl
Tf	triflyl, trifluoromethanesulfonyl, $CF_3S(O)_2$
THF	tetrahydrofuran; oxolane
TIPS	triisopropylsilyl
TMS	trimethylsilyl

1. Introduction

Glycoconjugates, which are biomolecules in which glycan chains are linked to lipids or proteins, play roles in various biological phenomena, including proliferation, differentiation, and motility of cells.[1] Furthermore, glycoconjugates exposed at the surface of bacteria act as ligands for innate immune receptors and activate innate immunity to induce inflammatory responses and production of cytokines.[2] Therefore, glycoconjugates expressed not only in humans, but also in various organisms are of considerable interest, and many synthetic studies of natural-type complex glycoconjugates have been reported. The role of synthetic chemistry in the preparation of glyco-conjugates with specific structures will remain important because of the diversity of lipid length in glycolipids and the heterogeneity of glycans in glycoproteins.[3]

Functional analysis of glycoconjugates is also an active area of research. The modes of action of glycoconjugates are often complex and difficult to analyze. However, functional analysis of natural products and small mole-cules has identified the target proteins of many compounds.[4] In the case of glycoconjugates, however, analysis mainly relies on genetic engineering and molecular biology methods, and complementary chemical biology approaches at the cellular level are quite limited, mainly due to the synthetic inaccessibility of suitable probe molecules, as well as the degradation of glycoconjugates in cellulo or in vivo. Depending on the cell type and the target glycan structure, glycoside hydrolases can cleave the terminal carbo-hydrate from glycoconjugate probes, which may alter their biological activ-ity, thus complicating functional analysis. The author's group is working on the development of pseudo-glycoconjugates, which are structurally modi-fied glycoconjugate analogs that may retain the original function of the parent molecule in cellulo/in vivo or acquire a new function. This article summa-rizes our recent work on the development of pseudo-glycoconjugates with a glycoside hydrolase-resistant C-glycoside linkage.

2. Design of C-glycoside analogs of gangliosides: Advantages of C-linked glycans

We have been interested in the relationship between gangliosides **1**, which are sialic acid-containing glycolipids, and their degrading enzymes, sialidases.[5] We have focused on the creation of ganglioside analogs (pseudo-gangliosides) that are not cleaved by sialidases. A sialidase normally trims the terminal (non-reducing end) sialic acids of gangliosides, and three main strategies have been proposed to provide resistance to sialidases[6] (Fig. 1): (**A**) replacing the oxygen atom of the O-glycosidic (O-sialoside) bond of sialic acid with another atom such as sulfur (S-sialoside) or carbon (C-sialoside), (**B**) introducing an F atom at the 3-position of the sialic acid, and (**C**) replacing the oxygen atom in the sialic acid ring with a carbon atom (carbasugar analogs). However, the synthesis of ganglioside analogs based on the approach **C** is difficult.[7] Approach **B** seems promising, and a disaccharide with an F atom at the 3-position of sialic acid has recently been reported.[8] However, we chose to focus on convergent synthesis using approach **A**.

In 1989, Hasegawa and Kiso reported the S-linked ganglioside GM4 analog **2**[9] (Fig. 2), which has similar biological activity to native GM4.[10] However, the S-sialoside bond differs significantly from that of the original O-sialoside bond in terms of bond length (C–O: 1.41 Å; C–S: 1.80 Å) and bond angle (dimethyl ether C–O–C: 112°; dimethyl sulfide C–S–C: 98.8°).[11] Thus, even if **1** exhibits sialidase resistance, it is conceptually somewhat distinct from a pseudo-ganglioside. From the viewpoint of structural similarity, a C-sialoside should be optimal (C–C: 1.53 Å; C–C–C: 113.8°), and furthermore, the introduction of a substituent into the C-sialoside bond would have the advantage of adding further functionalities that would not be accessible with S-sialoside. Initially, we thought that the greatest disadvantage of converting an O-sialoside to a simple CH_2-sialoside would be the decrease in the acidity of the sialic acid (calculated pK_a: from 2.31 to 3.34). Therefore, we thought that using a CF_2 group (calculated pK_a: 2.18) instead of a CH_2 group would be better. In 2007, we developed the

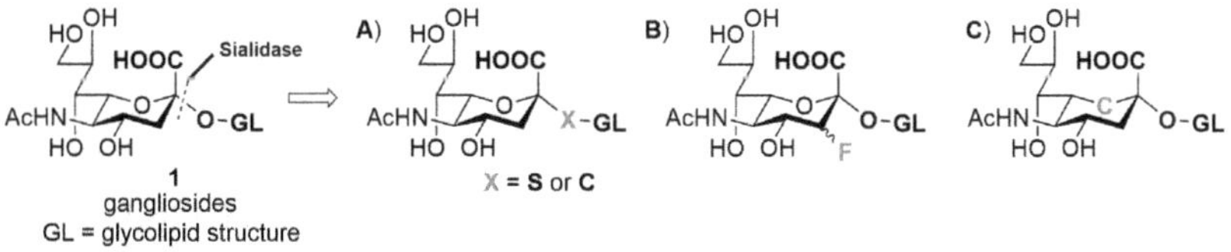

Fig. 1 General structure of gangliosides and three representative strategies for building sialidase-resistant analogs of sialo-glycoconjugates.

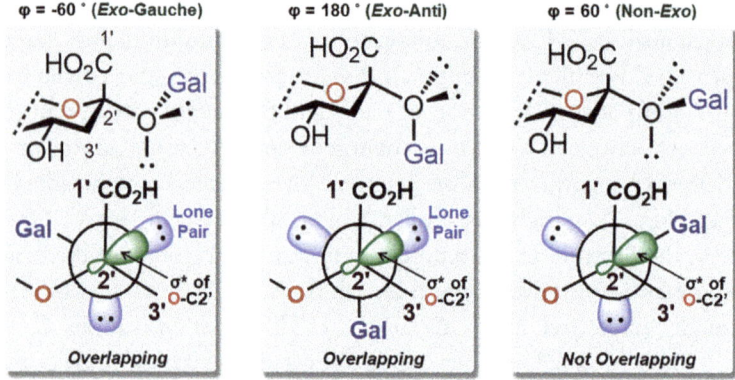

Fig. 2 S-linked and CF$_2$-linked GM4 analogs (**2** and **3**) and C-linked GM3 analogs (**4–7**).

X = Y = F: CF$_2$-linked GM3 (**4**)   X = H; Y = F: (S)-CHF-linked GM3 (**6**)
X = F; Y = H: (R)-CHF-linked GM3 (**5**)   X = Y = H: CH$_2$-linked GM3 (**7**)

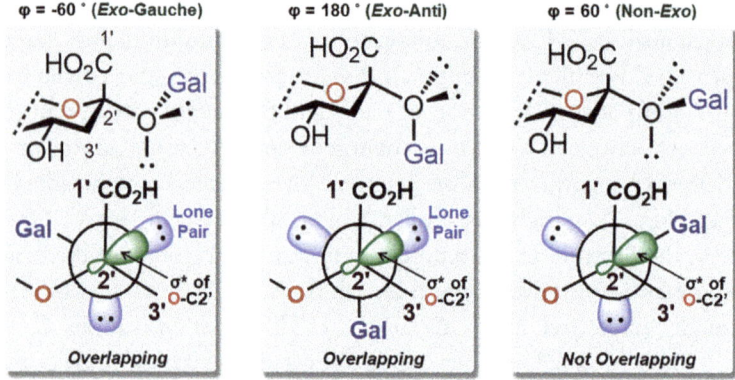

Fig. 3 Rotation of O-sialoside linkage and exo-anomeric conformations.

CF$_2$-linked GM4 analog **3** (Fig. 2) and confirmed that its biological activity is almost equivalent to that of native GM4.[12] This was the first example of the synthesis of a glycoconjugate analog with a C-glycosidic linkage in the glycan chain. We decided to expand this concept to the more biologically important GM3 analog **4**, but we had some concern about the validity of the CF$_2$-linked analog as a pseudo-ganglioside, due to the issue of glycan conformation.

As shown in Fig. 3, the exo-gauche and exo-anti conformations of the native O-sialoside bond are usually considered stable (natural-type conformations) due to a stereoelectronic effect known as the exo-anomeric effect. These conformations are favored in solution, as demonstrated by NMR and MD calculations.[13] X-ray crystal structure analysis of sialic acid-binding proteins has also shown that the proteins recognize sialo-glycans with either conformation.[14] Interestingly, a non–exo conformer of sialo–glycan can also be observed in the crystal structures, but does not participate in protein binding (Fig. 4).[15] This suggests that although the O–sialoside bond is rotatable, the conformations stabilized by the exo-anomeric effect are crucial for expression of the biological functions of sialo–glycans.

In the CF_2-linked analog **4**, however, the corresponding natural-type conformation should be destabilized by the unique stereoelectronic effect of the C—F bond, the so-called gauche effect (Fig. 5). Although the acidity problem could be overcome by using the CF_2-sialoside linkage, the CF_2-linked analog

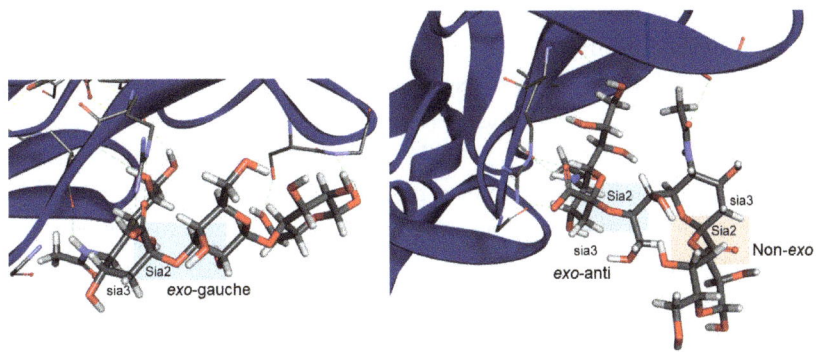

Fig. 4 Binding mode of sialo-glycans with reovirus attachment protein σ1.

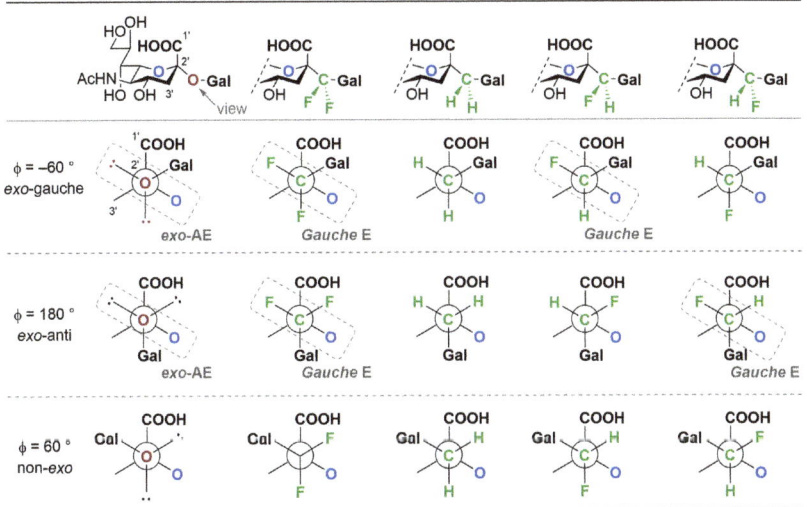

Fig. 5 Staggered conformations of O- and C-sialosides; exo-AE: exo-anomeric effect, Gauche E: gauche effect.

might not readily adopt natural-type conformations. Therefore, we turned our attention to the CHF-linked analogs (**5** and **6**, Fig. 2) with a single F atom. The stereochemistry of the CHF group influences the conformation of the C-sialoside bond, and the exo-anti conformation in the R-isomer and the exo-gauche conformation in the S-isomer become acceptable conformations due to the absence of one of the F atoms of the CF_2-linked analog **4** (Fig. 5). The acidity of the CHF-linked analogs is expected to be slightly lower than that of the natural molecule, but we considered that this would be in an acceptable range. The CH_2-linked analog **7**, on the other hand, is not electronically stabilized in any of the conformations because of the absence of stereo-electronic effects, as indicated by MD calculations.[13] The contribution of non-exo conformations would be limited due to the steric repulsion between the galactose unit and C-3 of the sialic acid moiety.

Overall, we expected that the CHF-linked analogs (**5** and **6**) would be the most suitable pseudo- sialoglycan analogs. As described above, C-glycosides not only confer metabolic tolerance, but also enable control over molecular properties such as acidity and conformation, which is difficult to achieve with approaches **B** and **C** (Fig. 1). To validate this concept, we set out to synthesize all four species of the GM3 analogs (**4–7**) and evaluate their biological activities.

3. Stereoselective synthesis of fluorine-containing C-sialosides

Various methods for constructing C-glycosidic linkages have been reported[16,17] However, most of them are focused on C-aryl glycosides with an aromatic ring connected to the anomeric positions and thus are monosaccharide analogs. Few C-glycoside analogs of glycan chains (in naturally occurring form) have been synthesized, though several glycoconjugates with monosaccharide structures, including a CH_2-galactosylceramide (KRN7000) analog, have been reported.[18,19] Several F-containing disaccharide-like C-glycoside analogs have also been constructed (Fig. 6),[20] but no analogs of native glycans had been reported.[21]

As mentioned earlier, we had already synthesized CF_2-linked GM4 (**3**). The key CF_2-sialoside bond was stereoselectively constructed by utilizing the Ireland–Claisen rearrangement reaction (Scheme 1).[12,22] Namely, the ester F_2-**10** was prepared as the precursor for the rearrangement reaction by linking the reduced sialic acid derivative **8** and the galactose derivative F_2-**9** modified with a difluoromethylene group at the C-3 position.

This reaction was then followed by treatment with excess LiHMDS and TMSCl at low temperature and warming to room temperature. Treatment of the resulting product with TMS-diazomethane gave the CF_2-sialoside α-F_2-11 in an α-selective manner. Although other methodologies for C-sialoside formation have also been reported, this strategy is currently the only method available to introduce F atoms into a C-sialoside.

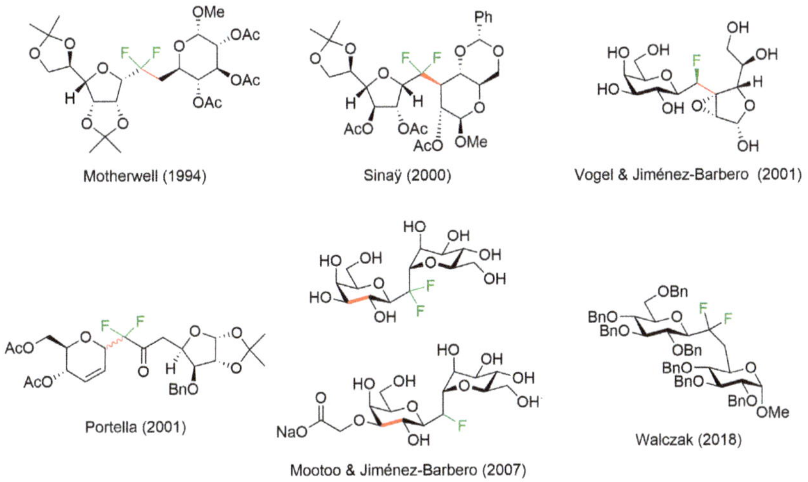

Fig. 6 Reported CHF- or CF_2-linked disaccharide-like compounds.

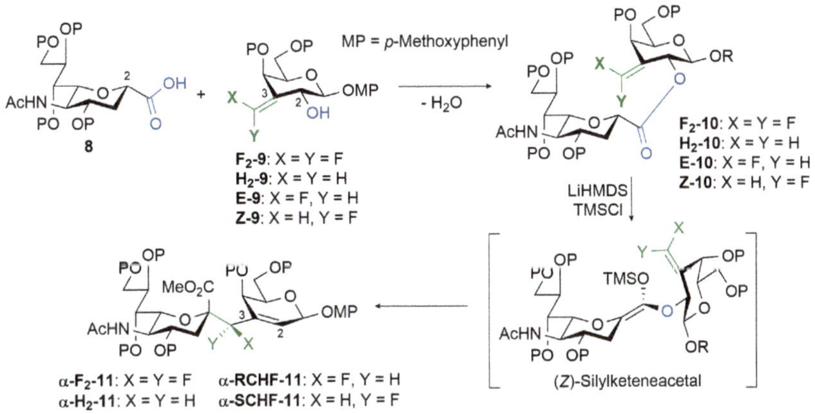

Scheme 1 C-sialoside construction by Ireland–Claisen rearrangement.

Therefore, we planned to utilize this method to construct CHF-sialosides. In order to selectively synthesize (R)- and (S)-CHF-sialosides, respectively, selective synthesis of precursors **E-10** and **Z-10**, which correspond to galactose derivatives **E-9** and **Z-9**, was thought to be necessary, considering the stereospecificity of the rearrangement reaction (Scheme 1). Although few methods are available for selective synthesis of a terminal monofluoromethylene group,[23] we thought that this could be achieved by utilizing the unique properties of the galactose skeleton, which we had discovered while synthesizing the CF_2-sialoside. The precursor of these galactose derivatives, 3-keto **12-TIPS**, was found to adopt a chair conformation, as judged from the J-value between H-1 and H-2, whereas the galactose derivative **F$_2$-9** with the difluoromethylene group is considered to adopt a boat-type conformation (Scheme 2). The J-value for the corresponding simple methylene derivative **H$_2$-9** suggested a similar chair confirmation to the 3-keto **12-TIPS**. This implies that the presence of an F atom in the exomethylene group causes allylic strain[24] with the bulky protecting group on O-2. Therefore, we expected that fluoromethylenation of the ketone **12-TIPS** would selectively afford an **E-9**-type compound due to steric repulsion between the F atom and the protecting group at O-2 in the transition state.

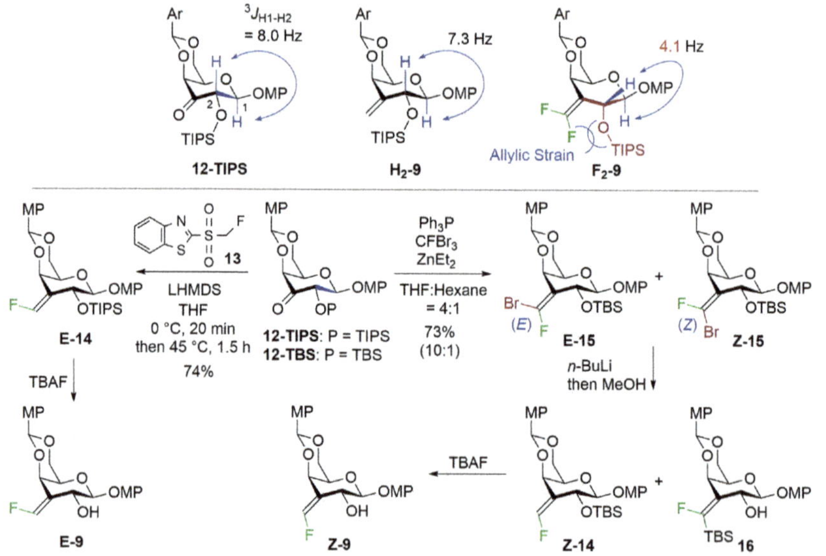

Scheme 2 Stereoselective synthesis of galactose derivatives **E-9** and **Z-9** with monofluoromethylene functionality at the C-3 position.

We performed a Julia–Kocienski-type reaction with the BT sulfone derivative **13**[25] and found that the desired **E-14** was selectively formed (Scheme 2) as we had expected. The selective synthesis of **Z-14** seemed difficult, but by using bromofluoroolefination,[26] we expected that **E-15** would be selectively formed due to steric repulsion with the Br atom, which is larger than the F atom. In fact, **E-15** was obtained in high yield with high selectivity when we employed $CFBr_3$, PPh_3, and Et_2Zn. In this case, the TBS group was the optimal protecting group at the 2-position, and the concentration and solvent composition had a significant effect on the reaction efficiency. **Z-14** was synthesized by Br—Li exchange followed by protonation at low temperature. Under these reaction conditions, the minor isomer **Z-15** gave vinylsilane **16**, which facilitated separation of the isomers.

After removal of the silyl group, condensation of the resulting galactose derivative (**E-9** and **Z-9**) with the sialic acid derivative **8** ($P = CH_2OBn$) gave the esters **E-10** and **Z-10**, respectively. The esters were treated with excess LiHMDS and TMSCl at low temperature and then heated. The resulting crude products were treated with TMS-diazomethane to afford the corresponding methyl esters **11**. Scheme 3 and Table 1 show the results obtained in the Ireland–Claisen rearrangement reaction from ester **E-10** and **Z-10** as well as F_2-**10** and H_2-**10**. The rearrangement proceeded smoothly from F_2-**10** at room temperature, giving only α-F_2-**11** in high yield.[12] On the other hand, in the reaction of H_2-**10**, the rearrangement also proceeded at room temperature, and although the desired α-H_2-**11** was obtained preferentially, the selectivity was poor (5:1).[1,5] Interestingly, the stereoselectivity decreased to 1.8:1 when the reaction temperature was lowered, and conversely improved to 10:1 when the mixture was refluxed in THF.[27]

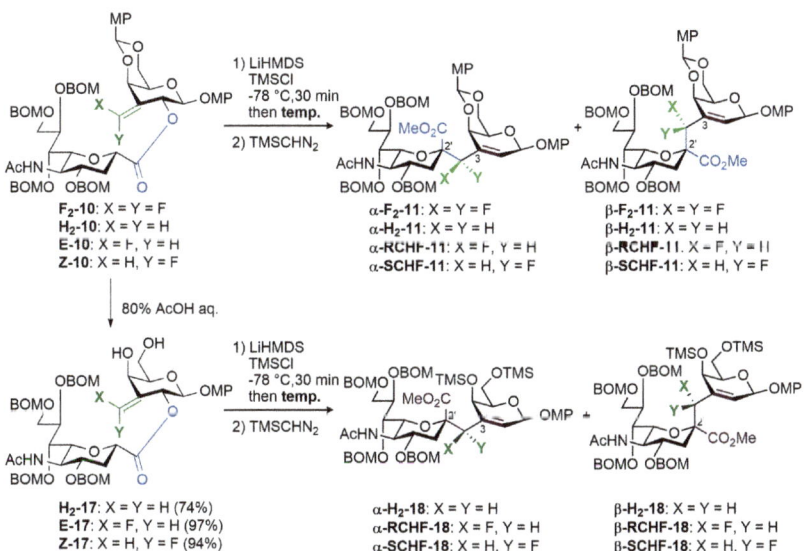

Scheme 3 Ireland-Claisen rearrangement reaction of ester **10** or **17**.

Table 1 Temperature dependence of the stereoselectivity in Ireland-Claisen rearrangement reaction of ester **10** or **17**.

Entry	Substrate	Temperature (°C)	Yield (%)	α:β ratio
1	F_2-**10**	rt	86	>20:1
2	H_2-**10**[a]	−20	25	1.8:1
3	H_2-**10**[a]	25	73	5:1
4	H_2-**10**	reflux	86	10:1
5	H_2-**17**	reflux	82	15:1
6	**E-10**	rt	82	3:1
7	**E-10**	reflux	78	4:1
8	**E-17**	reflux	92	>20:1
9	**Z-10**	rt	82	10:1
10	**Z-10**	reflux	81	8:1
11	**Z-17**[b]	rt	97	7:1
12	**Z-17**[b]	reflux	92	6:1

[a]The corresponding H_2-**10** with a benzylidene acetal group was used.
[b]The reaction time for condition 1 was 45 min.

A reasonable explanation for these results is steric repulsion between the substituents at the C-2 position and substituent **Y** on the olefin facing the C-2 position (Fig. 7). When the C-2 substituent is placed in the equatorial position, the galactose structure can maintain the chair conformation, but the transition state will be half-chair conformation **A**, leading to the desired **α-11**, while the twist-boat transition state **C** gives **β-11**. If the substituent **Y** is an F-atom, 1,3-allylic strain, as described above, directs the C-2 substituent to the pseudo-axial position, and facilitates access to chair-like transition state **B** to give the desired **α-11**. The temperature-dependent stereoselectivity observed when the substituent Y is an H-atom can be interpreted as an increase in the probability of rearrangement reaction via transition state **B** by accelerating the rate of conformational change of the galactose skeleton upon heating.

The validity of these considerations was supported by the outcomes of rearrangement of **E-10** and **Z-10**. Reaction of **E-10** proceeded at room temperature to give **α-RCHF-11**, but with a selectivity of 3:1 (Table 1, entry 6). Refluxing only improved the selectivity to 4:1 (Entry 7).

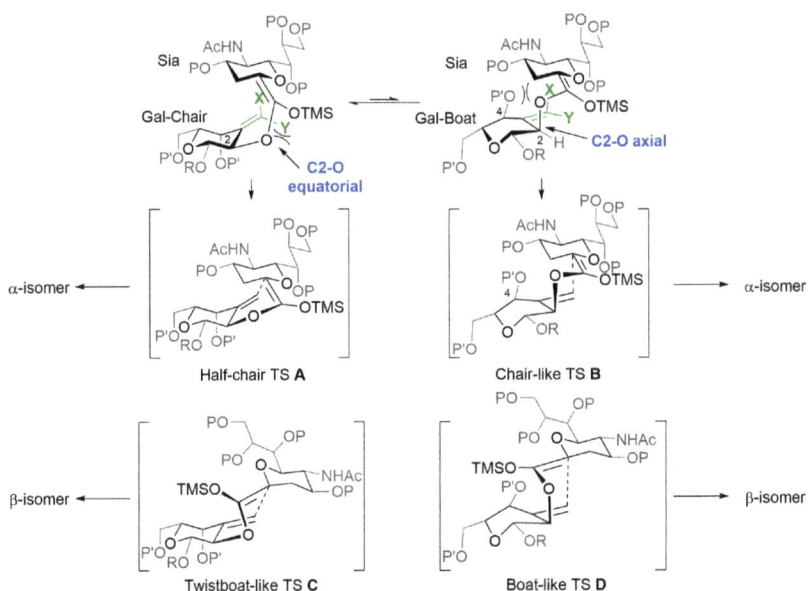

Fig. 7 Plausible transition state structure **A–D** in the Ireland–Claisen rearrangement reaction of **10** and **17**. Substituents **X** and **Y** were omitted in the transition state structures **A–D**.

The low selectivity can be attributed to the 1,3-allylic strain between the C-4 substituent and substituent **X** (F-atom) in the boat conformation of the galactose skeleton required for transition state **B**, preventing the conformational change of the galactose skeleton. On the other hand, in the rearrangement reaction of **Z-10**, the desired stereoselectivity for **α-SCHF-11** was observed even at room temperature (10,1, entry 9), and on heating, the selectivity conversely decreased to 8:1. As with **F₂-10**, the contribution of transition state **B** would be high even at room temperature due to the F-atom on the olefin (substituent **Y**), and heating promotes the unfavorable transition states **A** and **C**.

Then, removal of the *p*-methoxybenzylidene acetal group of the rearrangement product was found to be feasible only for **α-F₂-11**, while for other derivatives (**α-H₂-11**, **α-RCHF-11**, and **α-SCHF-11**), the acid-labile allyl acetal structure resulted in degradation of the galactose skeleton.[27] Therefore, we decided to perform the rearrangement reaction after removal of the *p*-methoxybenzylidene acetal. Without cyclic protection, the galactose skeleton would be more susceptible to conformational change, and the contribution of transition state **B** was expected to be higher. Indeed, the reactions of **CH$_2$-17** and **E-17** gave the desired **α-H$_2$-18** in 15:1 ratio and **α-RCHF-18** with perfect stereoselectivity, respectively, affording better results than were obtained with the precursors having the cyclic protective group. Conversely, the reaction from the **Z-17** showed slightly lower selectivity for **α-SCHF-18**. We speculate that this result, which is difficult to explain, might be related to the conformational flexibility of the galactose skeleton.

The resulting rearrangement product required the re-introduction of the 2-hydroxyl group and the 3-hydrogen atom on the galactose ring to the sialylgalactose structure. This was expected to be a simple transformation, since there is a double bond at the C-2–C-3 position. However, the trisubstituted C-2–C-3 double bond with many peripheral functional groups was found to be inert under the reaction conditions we investigated. Eventually, a seven-step synthetic sequence was established. As shown in Scheme 4, all of the rearrangement products **α-F$_2$-11**, **α-H$_2$-18**, **α-RCHF-18**, and **α-SCHF-18** were converted to the lactone **19**, which enabled stereoselective dihydroxylation of the C-2–C-3 double bond to introduce the 2-OH group, affording **20**. The 3-OH group was then removed by radical-mediated reduction via the corresponding cyclic thiocarbonates to obtain the C-sialoside analogs **21** with the 2,3-sialylgalactose structure. Although the seven-step transformation was applicable for preparation of all the C-sialoside analogs we designed, a difference was observed in the hydrolysis of the methyl ester of the sialic acid moiety. The reaction is a simple hydrolysis reaction, but the product decomposed at higher temperatures or during prolonged reaction of the CHF-linked analogs. In contrast, no such degradation reaction was observed for the CF$_2$-linked and CH$_2$-linked analogs. In addition, as mentioned earlier, the resulting carboxylic acids were problematic because of the instability of the allyl acetal structure under acidic conditions. Thus, in the case of the most unstable CH$_2$-linked compounds, EDC and DMAP were added directly to the dichloromethane solution of the crude material before concentration, and the solution was concentrated during lactoniziation to prevent degradation.

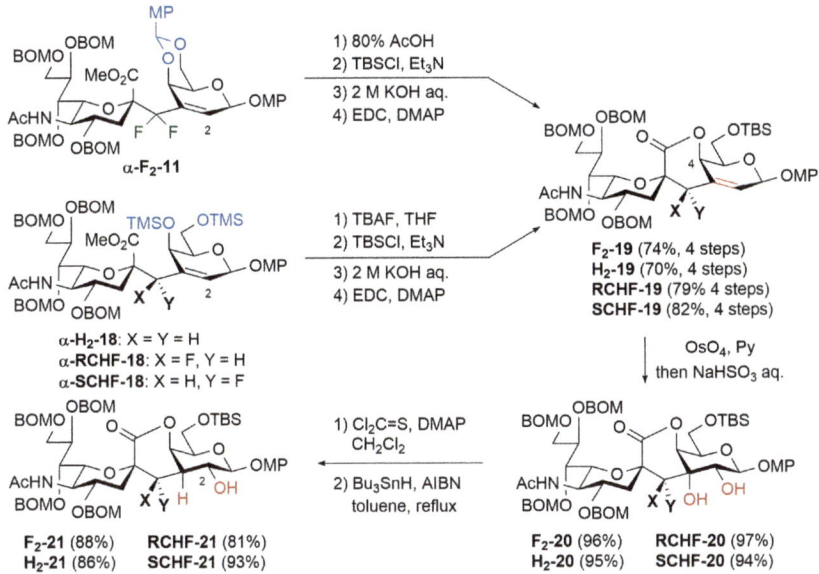

Scheme 4 Transformation to sialylgalactose analogs **21** with a C-sialoside linkage from the corresponding rearrangement products.

4. Unique reactivity of sialylgalactose donors with fluorine-containing C-sialoside bonds and synthesis of GM3 analogs

With the disaccharide **21** with C-sialoside linkages in hand, the remaining challenge in synthesizing the GM3 analogs was the glycosylation reaction with the common glucosylceramide unit. The previous CF_2-linked GM4 analog **3** was synthesized by means of a similar glycosylation reaction. But, we were left with somewhat of a challenge. The C-2 hydroxyl group of the CF_2-linked disaccharide F_2-**21** was protected with an acetyl group, which can be involved in neighboring-group participation during the β-selective glycosylation reaction, and the C-1 functionality was changed to trichloroacetoimidate to prepare the donor F_2-**22** (Scheme 5-1). Glycosylation with the protected ceramide **23** gave the desired F_2-**24** in only 20% yield with 40% recovery of lactol derivative F_2-**25**.[12] This result suggests low donor reactivity of F_2-**22**.

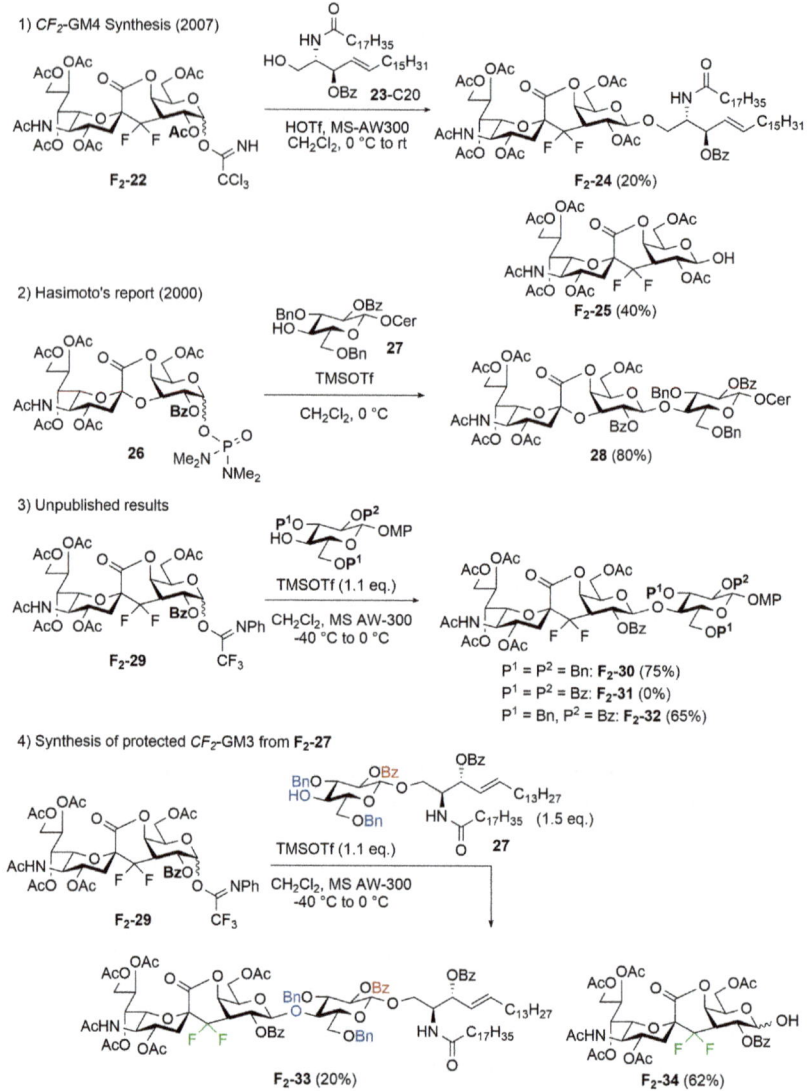

Scheme 5 Several attempts at the glycosylation of CF$_2$-linked sialylgalactose donors with a lactone ring.

On the other hand, synthesis of native GM3 by glycosylation of natural-type donor **26**, which has a similar lactone structure, with glucosylceramide derivative **27** was reported by Hashimoto and co-workers.[28] Although the leaving groups are different, they found that GM3 precursor **28** was obtained in high yield (Scheme 5-2). Comparing these results, we first decided to change the protecting group of the C-2 hydroxyl group to a Bz group.

Since a CF_2-linked donor such as F_2-**22** has many electron-withdrawing groups as well as the CF_2-group, the oxonium intermediate generated in the glycosylation reaction would not be stabilized by the substituents on the donor molecules. In this case, an intermediate with a counter anion (such as OTf) attached directly to the anomeric position rather than the oxonium species would be the dominant species, and some examples have been reported in which the reactivity of the acceptor as a nucleophile determines the success or failure of the glycosylation reaction for such intermediates.[29] Therefore, glycosylation of donor F_2-**29** with glucose acceptors having different protecting groups was investigated (Scheme 5-3). Instead of trichloroacetoimidate, we tried the more stable N-Ph- trifluoroacetoimidate[30] as the leaving group.

The acceptor in which all hydroxyl groups on glucose were protected with electron-withdrawing Bz groups gave no product F_2-**31**, but changing two of them to Bn groups gave F_2-**32** in 65% yield. When Bn protecting groups were used for all hydroxyl groups, the yield of product F_2-**30** improved to 75%, indicating that the reactivity of the acceptor was important for the glycosylation of F_2-**29**, and that the use of electron-withdrawing protecting groups should be avoided as far as possible.

Therefore, glucosylceramide derivatives **27** with two Bn groups was prepared. This is the same compound used in Hashimoto's report. Glycosylation was performed under the same conditions as before, but again, only 20% of the target product F_2-**33** was produced, and 60% of lactol F_2-**34** was recovered (Scheme 5-4). These results indicate that replacement of the O-sialoside bond with the CF_2-sialoside bond significantly reduced the reactivity of the donor.[31] The chemical yield of this glycosylation did not improve despite multiple attempts using various amounts of the acceptor or Lewis acid. The fact that the donor was recovered in the form of lactol F_2-**34** confirmed that the donor was activated, but the acceptor, glucosylceramide derivative **27**, was barely recovered, suggesting that it might have been degraded under Lewis acid conditions. The product F_2-**33** also contained impurities that appeared to be degradation products of the ceramide part, making complete purification difficult. These results led us to believe that the product F_2-**33** decomposed under the strong Lewis acidic conditions, and that activation under milder conditions would be essential to obtain F_2-**33** in high yield and purity.

Therefore, we decided to use the glycosylation reaction promoted by the π-acidic Au(I) complex, reported by Yu and co-workers.[32,33] We found that the results in the synthesis of glucosylceramide derivatives performed under Lewis acidic and Au-catalytic conditions were dramatically different.

We had synthesized the glucosylceramide derivatives by glycosylation of the
N-Ph trifluoroacetimidate donor **35** with the ceramide **23–C18**, but despite
extensive trials, the yield was only moderate (Scheme 6-1). Again, there was
a little recovery of **23–C18**, and it seemed that decomposition of the product
gradually occurred during consumption of the starting material. On the other
hand, glycosylation of alkynylbenzoate donor **37** with **23–C18** promoted by
PPh₃AuNTf₂ cleanly gave **38** in quantitative yield (Scheme 6-2). The suppression of degradation of the ceramide moiety by strong Lewis acidic
conditions enabled the reaction to be carried out at room temperature, overcoming the low solubility of ceramide **23–C18** at low temperature, and this
would have contributed to the high yield under gold(I)-catalytic conditions.

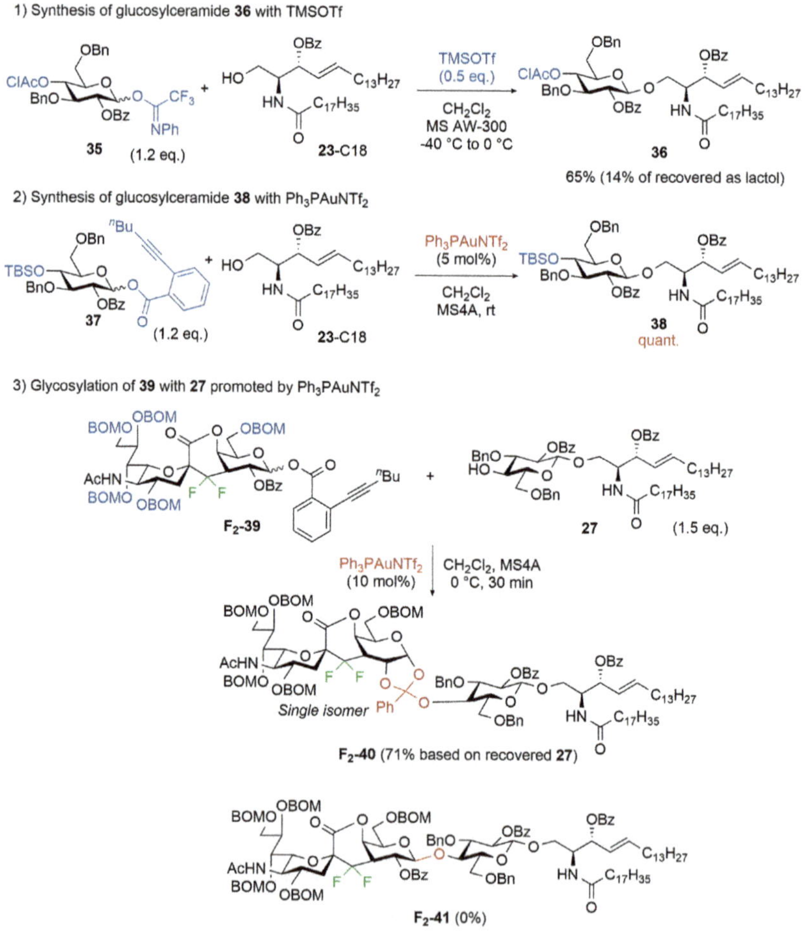

Scheme 6 Synthesis of glucosylceramide derivatives **36** and **38** and the glycosylation of CF₂-linked sialylgalactose donor **F₂-39** with **27** promoted by gold complex.

Based on this finding, donor F_2-39 was prepared, in which an alkynylbenzoate group was also introduced into CF_2-linked sialylgalactose. Previously, all hydroxyl groups were protected with acyl groups to obtain resistance to Lewis acid conditions, but under gold complex conditions, it should be possible to use acetal protecting groups, which are not available in normal glycosylation reactions. Therefore, considering the increase of the stability of the oxonium ion resulting from the absence of the electron-withdrawing acyl group, we synthesized donor F_2-39, in which all the hydroxyl groups except the 2-position Bz group were changed to $BnOCH_2$ (BOM) groups. The glycosylation of F_2-39 with 27 in the presence of PPh_3AuNTf_2 proceeded smoothly to provide the adduct in 70% yield. However, this proved to be orthoester F_2-40, and the expected glycoside F_2-41 was not formed at all (Scheme 6-3). These results indicated that, although the glycosylation with gold complexes achieved an improvement in the mass balance, our expectation regarding the reactivity of the donor F_2-39 was incorrect.

During β-selective glycosylation utilizing neighboring-group participation by a C-2 acyl group, orthoesters are often formed. The orthoester is generally re-activated again under acidic conditions to generate cationic intermediates and gradually give a thermodynamically stable β-glycoside. Thus, the desired β-glycoside can be obtained by increasing the reaction temperature or prolonging the reaction time. On the other hand, the orthoester F_2-40 that is kinetically produced in our glycosylation reaction could not be re-activated due to the low oxophilic Lewis acidity of the gold(I) complex, and once formed, it can be considered to remain intact in the reaction mixture (Scheme 7). Since no β-glycoside F_2-41 was produced at all, the kinetic pathway to provide the β-glycoside F_2-41 was significantly disfavored in the glycosylation reaction of donor F_2-39, i.e., its activation energy is significantly greater than that of the process leading to orthoester F_2-40. This indicates that even if an orthoester like F_2-40 could be re-activated using TMSOTf, the reaction rate to β-glycoside F_2-33 would be significantly slower. Thus, we concluded that the decomposition of the glucosylceramide 27 and the formation of the product F_2-33 compete under strong Lewis acid conditions. Although β-glycoside F_2-33 was gradually produced, the yield was low, and a significant amount of lactol F_2-34 was recovered after work-up. We had thought that the neutral chemical species 42 and oxonium species 43 would be generated first by the activation of the donor, and that the bisoxonium ion 44, where the neighboring Bz-group interacts with the oxonium species, and the cation is delocalized to the benzoyl group, was unlikely to be involved in the reaction due to the distortion induced by the lactone ring. However, these results strongly suggested a major contribution of bisoxonium ion **44**, and we realized that our expectations here had been in error. The effect of the BOM group in this reaction was negligible, and the same result was obtained with the corresponding acetyl-protected donor. The only way to successfully proceed with this glycosylation was to find a donor that would give β-glycoside in a kinetically controlled manner.

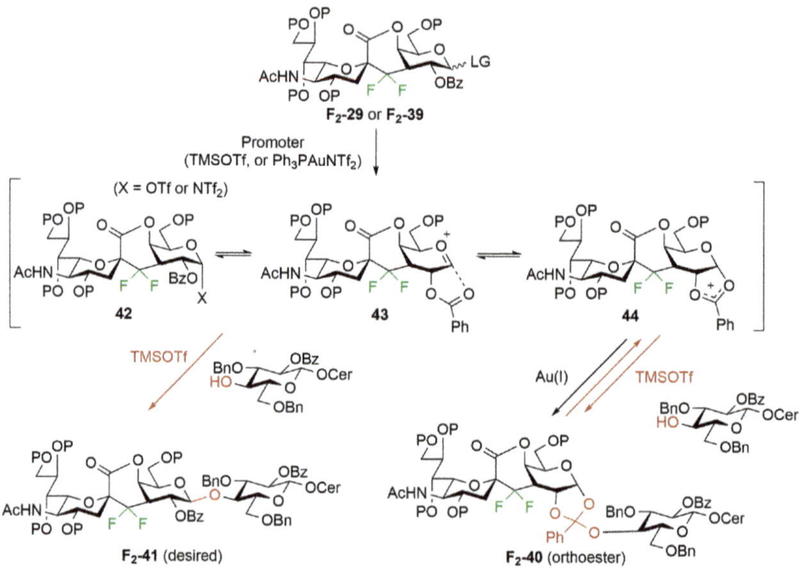

Scheme 7 Plausible intermediates in the glycosylation of F_2-**29** and F_2-**39**.

Although the concept of kinetically controlled glycosylation has not been explicitly discussed, there have been many studies of donor reactivity.[34] In general, reactivity in glycosylation is strongly influenced not only by the protecting group, but also by the conformation of the donor. Thus, simple cleavage of the lactone ring is likely to change the reactivity of the donor. It was not clear how much the kinetic pathway would change, but we decided to give it a try.

The synthesis of the lactone ring–opened donor is shown in Scheme 8. The C-2 hydroxyl group of F_2-**21** was protected with a Bz group, and then the TBS group was changed to a BOM group to give F_2-**45**. The lactone was cleaved with NaOMe and the resulting C-4 hydroxyl group on galactose was capped with an Ac group to afford F_2-**46**. Note that this step was conducted at low temperature because of the tendency to revert to the lactone F_2-45. After oxidative removal of the MP group at the reducing end, the donor F_2-47 was obtained by introducing alkynylbenzoate as a leaving group. The lactone ring, which was necessary for the introduction of the C-2 and C-3 substituents on the galactose ring, was eventually cleaved.

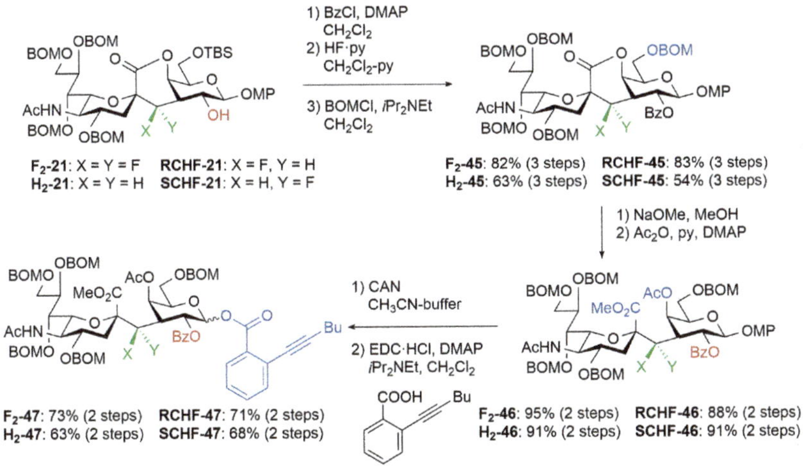

Scheme 8 Synthesis of ester-type donors **47**.

Glycosylation of the synthesized donor F_2-47 with the glucosylceramide derivative **27** promoted by the gold(I) complex successfully provided the desired β-glycoside F_2-48 in a high yield of 85% (Scheme 9). Interestingly, no orthoester F_2-49 was detected in this case. This result indicated reversal of the kinetic pathway, suggesting a drastic decrease of the activation energy for formation of β-glycoside F_2-47 in the absence of the lactone ring. Therefore, all C-sialoside analogs were converted to donors **47** similar to CF_2-linked F_2-47, and glycosylation with **27** was performed. Fortunately, β-glycosides **48** were similarly obtained as the main product in all cases, giving **H$_2$-48**, **R-CHF-48**, and **S-CHF-48** in 74%, 76%, and 59% yields, respectively. The formation of orthoesters was hardly observed, except in the case of the S-CHF linkage, where the orthoester **S-CHF-49** was formed in 20% yield. These results indicated that not only the presence or absence of the lactone ring, but also the presence and stereochemistry of the F atom affect the magnitude of the activation energy. Namely, the CF_2-linked and R-CHF-linked donors F_2-47 and **R-CHF-47** did not afford orthoesters, while only the S-CHF-linked donor **S-CHF-47** gave orthoesters. Although it was difficult to explain the formation of the orthoester **S-CHF-49** at this stage, the stereoelectronic effect of the F atom might affect the conformation and/or electronic interaction in the transition state.

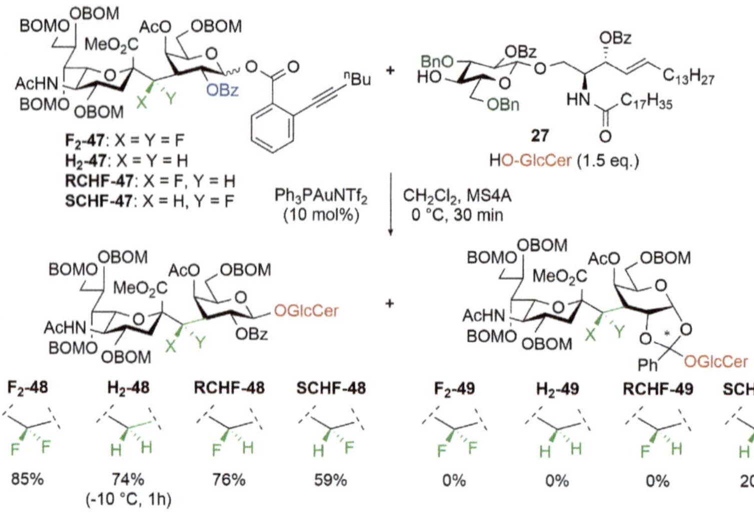

Scheme 9 Glycosylation of ester-type donors **47** with **27**.

The acyl groups of β-glycoside **48** were hydrolyzed under basic conditions, followed by hydrolysis of the methyl ester to give the corresponding carboxylic acids (Scheme 10). All remaining Bn or BOM groups were removed by Birch reduction to complete the synthesis of C-linked GM3 analogs.

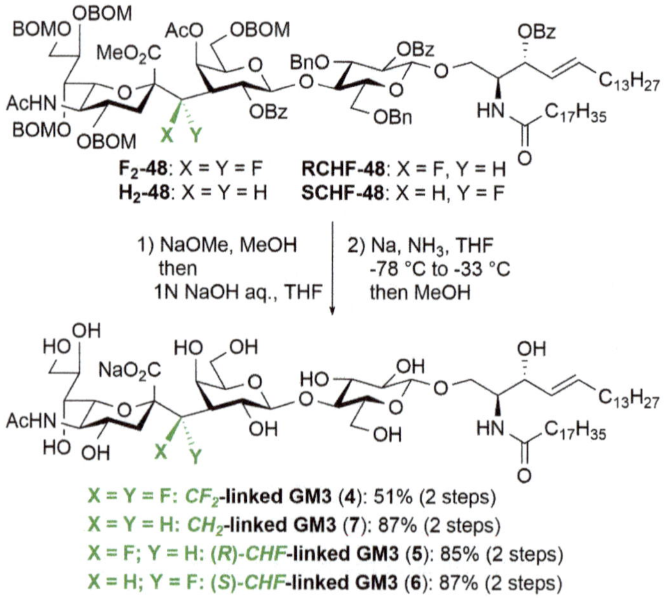

Scheme 10 Synthesis of C-linked ganglioside GM3 analogs (**4–7**).

5. Conformational analysis of CHF-linked sialylgalactoses

As mentioned earlier, the native sialylgalactose structure mainly adopts the exo-gauche and exo-anti conformations based on the dihedral angle φ (Fig. 5), and these structures have been confirmed by NOE correlations between the sialic acid and the galactose, and MD or MM calculations in most cases.[35] It was proposed that the dihedral angle ψ of the disaccharide with native O-sialoside lies in the range of −60 to +60° (Fig. 8). Using the same MD calculation method without any NMR data, Prof. Barbero proposed that a simple CH_2-linked disaccharide has flexible conformations in terms of both φ and ψ.[13]

On the other hand, in the case of C-sialoside analogs, it seems possible to perform conformational analysis based on NMR coupling constants (J-values).[36] The two dihedral angles involved in glycan conformations (φ for the rotation of the glycosidic bond and ψ for rotation of the bond next to it) affect the coupling constants for the surrounding elements. Since it is difficult to distinguish between the two prochiral F or H-atoms of the CF_2-linked or CH_2-linked analog, conformational analysis was investigated for both isomers of CHF-linked analogs. The dihedral angle φ can be estimated in terms of the J-values between the C atoms at positions 1 and 3 of sialic acid and the H- and F-atoms of CHF-sialoside, while the dihedral

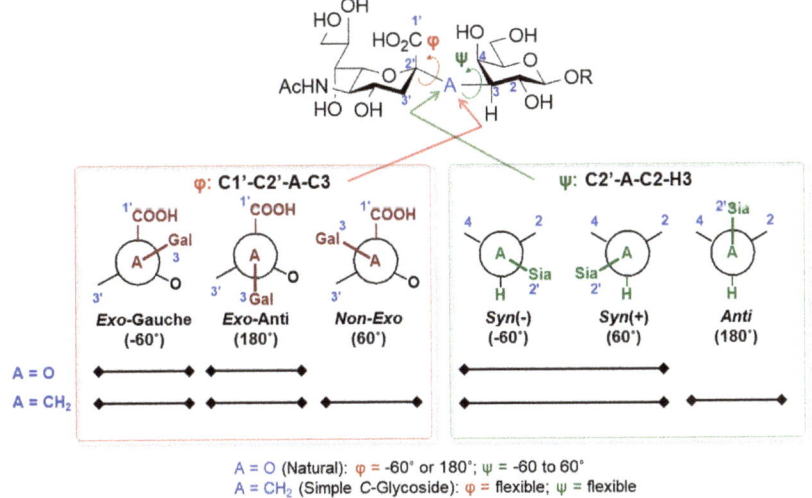

Fig. 8 Reported conformational properties of O-sialoside and CH_2-sialoside.

angle ψ can be estimated in terms of the J-values between the H and F atoms of CHF-sialoside and the C-atoms at C-2 and C-4 of galactose, as well as the C-2 atom of sialic acid and the H atom on C-3 of galactose. The ^3J-values were measured by the J-IMPEACH-BMC,[37] HETLOC,[38] or J-HMBC method.[39] On the other hand, DFT calculations were used to estimate the theoretical values. Although we initially considered estimating the stable conformation by DFT calculations, there were many factors to consider, such as the contribution of a countercation and the influence of solvents, so we decided to perform a conformational analysis using coupling constants, for which experimental values were available. It should be noted that the stereochemistry of CHF-sialoside was confirmed to be in accordance with the stereospecificity of the Ireland–Claisen rearrangement reaction based on the NOE correlations and coupling constants of the lactone-type compounds derived from compounds **21**.

We used a simplified DFT calculation procedure for this conformational analysis, as follows. First, considering the possible staggered conformations for φ and ψ, nine different conformations can be generated. These were set as initial structures, which were then optimized by means of PM3 level, DFT level (solvent-free), and HF (in water) calculations. It should be noted that the dihedral angle φ did not change much ($\pm 10°$) during the optimization process, but the dihedral angle ψ could change significantly. Without regard to which structure would ultimately be stable, the coupling constants for these optimized structures were estimated by means of DFT calculations (MPW1PW91/6-311 + G(d,p)).[40] If certain conformations are preferentially present in aqueous solution, the differences between the theoretical and measured values of all coupling constants related to the dihedral angle should be small for each φ and ψ. The differences between these theoretical and measured values are represented as bar graphs in Figs. 9 and 10. The discussion here is limited to a qualitative evaluation and does not extend to quantitative evaluation or to the rotational barrier energies between conformations.

Fig. 9 shows the observed J-values of the S-CHF-linked sialylgalactose derivative **49** measured on a 600-MHz instrument and the difference from the corresponding theoretical values. $^3J_{C-F}$ has a larger absolute value than $^3J_{C-H}$, so the errors are likely to be larger. Nevertheless, the bar graphs clearly indicate that **49** has an exo-gauche conformation ($\varphi = 68°$), and the contribution of non-exo conformation appears to be negligible. The errors in the coupling constants were large for dihedral angles close to any staggered conformation with respect to ψ, but good agreement was observed between the

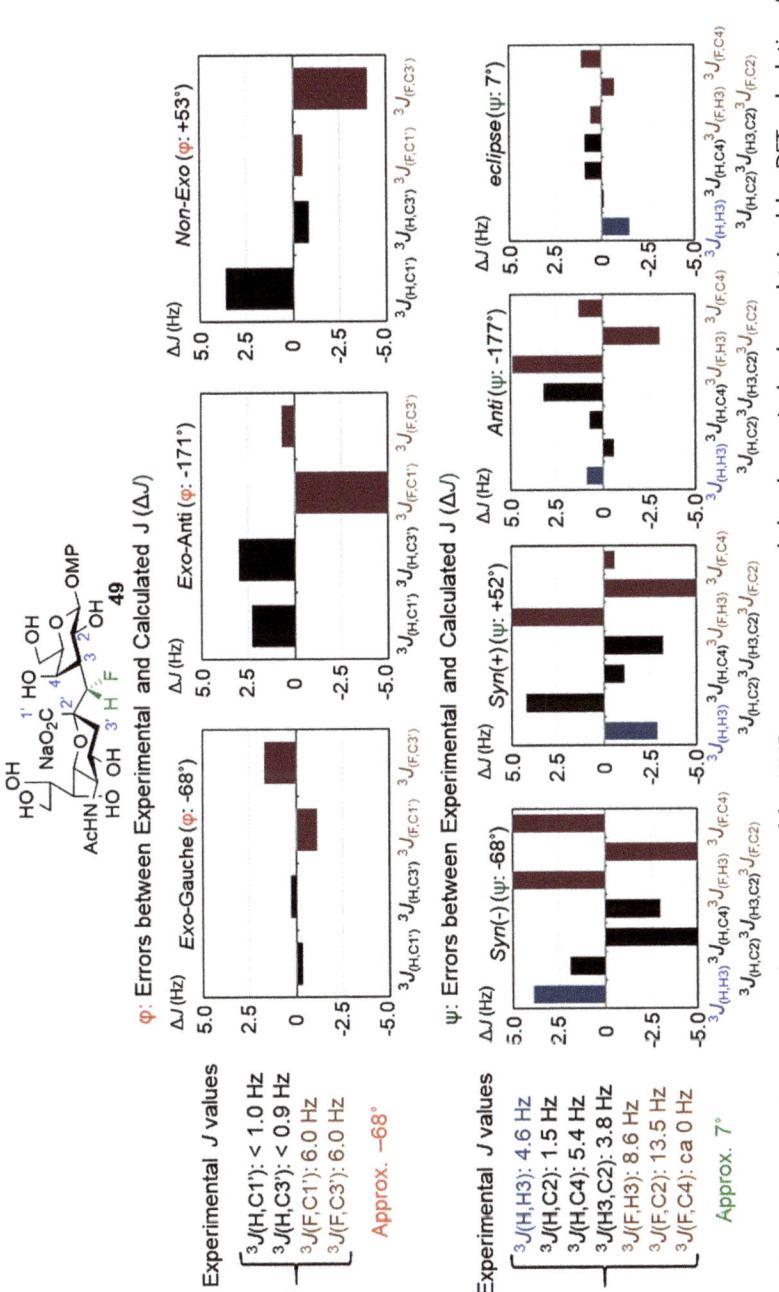

Fig. 9 Difference between actual J-values measured by NMR spectroscopy and the theoretical values obtained by DFT calculation for S-CHF-linked 2,3-sialylgalactose derivative **49**.

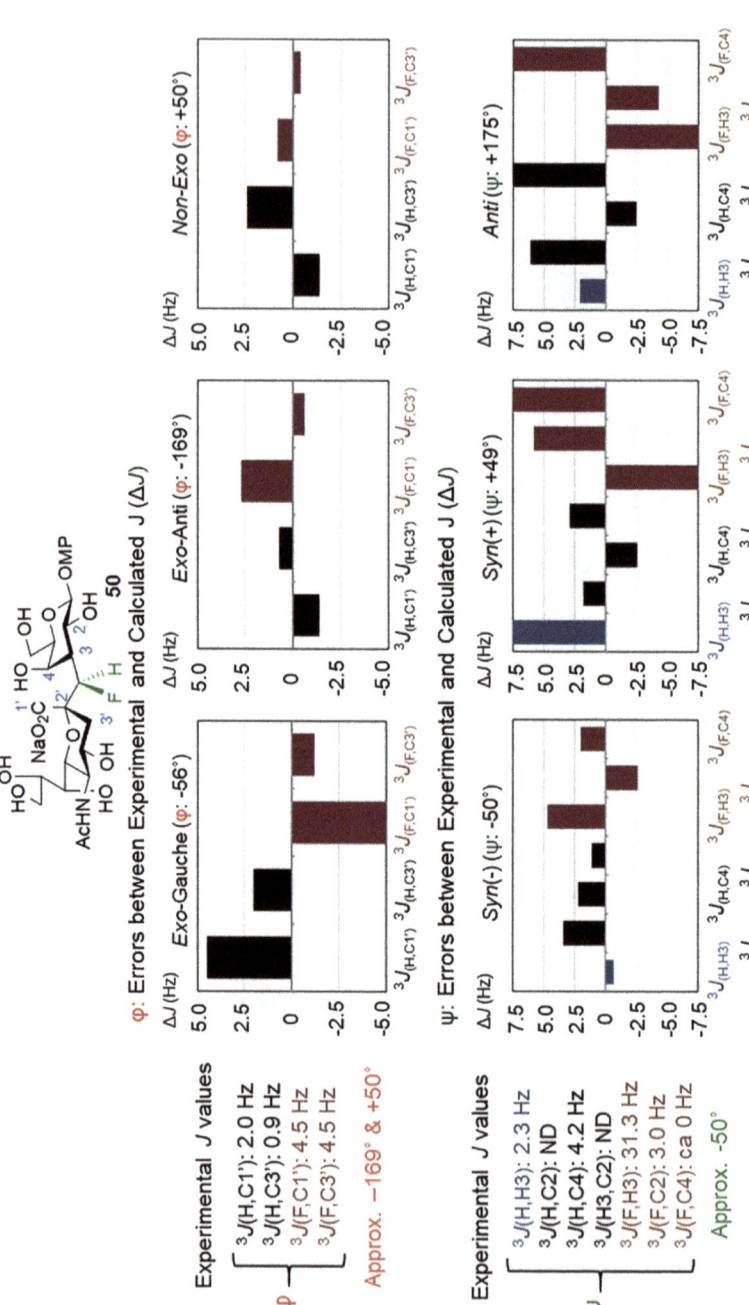

Fig. 10 Difference between the actual *J*-value measured by NMR spectroscopy and the theoretical value obtained by DFT calculation for R-CHF-linked 2,3-sialylgalactose derivative **50**.

theoretical and experimental values for the conformation corresponding to the eclipse conformation ($\psi = 7.2°$) found during the optimization process. Even in the natural O–sialoside of 2,3-sialylgalactose, ψ was speculated to adopt a similar synclinal conformation in the exo-gauche conformation, so the putative conformation was considered reasonable. We also found that the model using these two dihedral angles well explains the observed ROE correlations (Fig. 11). Taking these results together, we can conclude that the S-CHF-linked sialylgalactose derivative **49** adopts the exo-gauche conformation, with little contribution from the non-exo conformation.

A similar analysis was performed for the R-CHF-linked sialylgalactose derivative **50**, though this proved to be somewhat more complicated (Fig. 10). While the exo-gauche conformation contributes little, as expected from the gauche effect of the F atom, both the exo-anti and non-exo conformations showed better agreement with the theoretical values, suggesting that both conformations are present. On the other hand, although the errors were uniformly small for ψ in the syn(−) conformation, the involvement of multiple conformations for φ suggests that other conformations for ψ could not necessarily be ruled out. We found that models fixing the dihedral angle ψ at −50° and applying the exo-anti (−169°) or non-exo (50°) conformation for φ could explain the observed ROE correlations (Fig. 11). This strongly suggests the existence of two conformations, as expected, although it is not clear which is the major conformation. However, the major conformation can be determined from the sign of $^2J_{C-H}$ between the C atom at

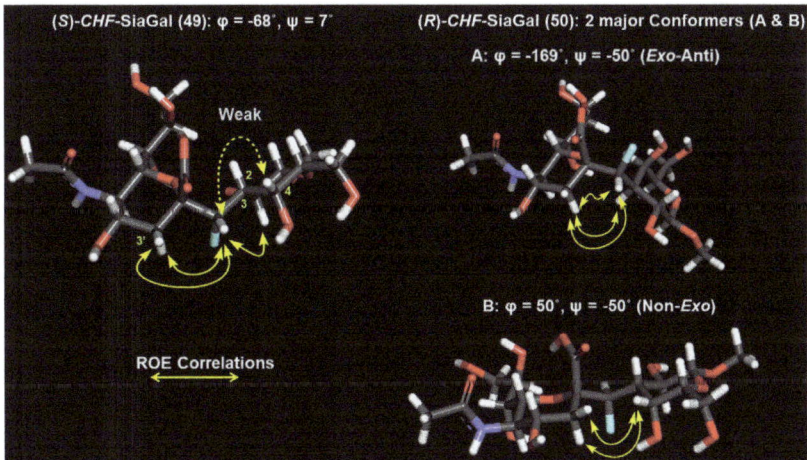

Fig. 11 Observed ROE correlations in the estimated conformations of CHF-linked sialylgalactose analogs (**49** and **50**).

the C-2 of the sialic acid and the H atom of the CHF group. The theoretical values are +1.9 Hz for the exo-anti ($-169°$) and -4.8 Hz for the non-exo ($50°$) conformation. The measured value was 3.0 Hz, but this was an absolute value and the sign was not determined. Since two conformations were found to exist, we did not attempt further analysis, but we believe the results show that this method is useful for qualitative conformational analysis of glycans with a C-glycosidic linkage.

6. Biological activity of C-linked pseudo-GM3s

Various biological functions have been reported for ganglioside GM3. Since it has a glycolipid structure, GM3 is likely to function on biological membranes, where it would interact with membrane proteins. Thus, for evaluating the biological activity of metabolically resistant analogs, examination at the cultured cell level would be preferable, and after performing various bioactivity studies, we concluded that quantitative evaluation of effects on cell proliferation would be consistent with our objectives. However, in measuring the effect of gangliosides on cell growth, it is important to consider the influence of the serum used, as pointed out by Ladisch and co-workers.[41] Therefore, we focused on assay systems in which cell proliferation could be achieved without the use of serum, and finally decided to evaluate the proliferation-promoting effect of GM3 on Had-1 cells. Taki and co-workers reported that the proliferation of Had-1 cells derived from murine breast cancer cells was markedly enhanced when GM3 was added exogenously, whereas it was suppressed by the addition of LacCer.[42] In this system, 2% fetal bovine serum (FBS) is used during passaging culture, but to examine the effect on proliferation, a supplement called HB-101 was added instead of FBS. Thus, this system should essentially eliminate the effect of serum. We first confirmed the growth-promoting effect of native GM3, and found that the number of cells was increased 1.66-fold compared to untreated cells on the fourth day of culture in the presence of 10 μM of GM3. Treatment with LacCer had no effect on proliferation in this system. When the four synthesized C-linked analogs were evaluated, CH_2-linked 7 and CF_2-linked 4 showed growth-promoting activity similar to that of native GM3 (1.51- and 1.62-fold, respectively). In contrast, CHF-linked analogs 5 and 6 showed higher growth-promoting activity, with a 1.85-fold increase in cell number for the R-CHF-linked analog 5 and a 2.04-fold increase for the S-CHF-linked analog 6 compared with native GM3.

This result suggested that the CHF-sialoside linkage more accurately reproduces the properties of O-sialosides, as we had hypothesized. The two CHF-sialoside analogs may have similar structures and physical properties, but their activities differ depending on the stereochemistry of the CHF group. Therefore, we speculated that the glycan conformation is important for this biological activity, and it is noteworthy that the S-CHF-linked analog fixed in the exo-gauche conformation exhibited higher biological activity.[43]

7. Direct C-glycosylation approach based on anomeric radical species

Although we have successfully synthesized GM3 analogs and established design guidelines for metabolically resistant analogs, we still had only one example of an analog with a C-glycoside linkage between two sugars of native glycoconjugates. Thus, we wanted to obtain further analogs to verify the usefulness of F-containing C-glycoside analogs. In addition, in order to efficiently utilize such analogs as biological tools or for innovative applications, it was necessary to streamline the synthetic methodologies. The author has been working on the development of more efficient synthetic methods for C-glycoside analogs since his transfer to Kyushu University. Some details of this early research will be described below.

Two main factors contribute to the complexity of C-glycoside synthesis: first, the requirement to achieve C-glycoside construction with reliable stereoselectivity, and second, the need for functionalization of the carbohydrate-ring after C-glycosidic linkage construction. The C-sialoside construction method developed by our group is still one of the most versatile methods to construct 2,3-sialylgalactose structures, but it employs an intramolecular reaction (Ireland–Claisen rearrangement), and only 2,3-sialylgalactose analogs can be synthesized. Since then, a number of C-glycosylation methods had been developed or are actively being studied,[16] but most of them target aryl C-glycosides[17] with aromatic compounds attached at the anomeric positions. In 2013, Werz and co-workers reported that C-glycosylation methods based on sp^2–sp^2 (or sp^2–sp) cross-coupling could be used to synthesize C-linked disaccharides (Scheme 11-1).[44] After cross-coupling, the double bond in the glycal moiety at the non-reducing end and the double (or triple) bond at the anomeric position are reduced or functionalized, leading to a variety of CH$_2$-linked disaccharide analogs. Although this method is useful for the synthesis of C-glycoside analogs, we wished to develop a more efficient direct C-glycosylation methodology.

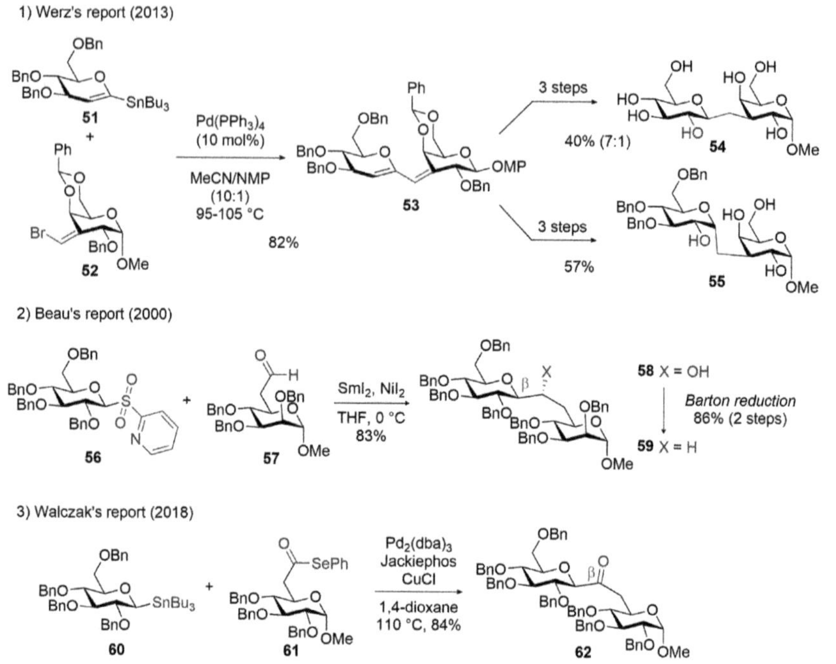

Scheme 11 Representative methods for the synthesis of C-linked disaccharides.

Native O-glycosides are generally synthesized by so-called glycosylation reaction, in which cationic intermediates generated by activation of donor molecules undergo nucleophilic attack by acceptors (Scheme 12-1). Various methods have been developed to control the stereochemistry at the resulting anomeric positions, and new catalytic methods are still being actively inves-tigated.[45] Here, the term "direct C-glycosylation" refers to a method in which the anomeric position of stable and storable donor molecules is acti-vated, as in O-glycosylation reactions, and the resulting intermediate can be connected to stable one-carbon-elongated acceptors through intermolecular reactions (Scheme 12-2). Although controlling the stereochemistry of the anomeric position requires a different strategy and is more challenging, we believed that we could overcome the drawbacks of the previous methods, which requires glycal derivatives as donor molecules, and thereby streamline the synthesis of C-glycoside analogs. A direct C-glycosylation method that can construct CH(OH)-glycosides using SmI_2-mediated Berbier-type (or Reformatsky-type) couplings has already been reported (Scheme 11-2).[46] This enables stereoselective C-glycosylation dependent on the 2-OH group, but there are difficulties in the conversion of CH (OH)-glycosides. Just before we started our research, Walczak's group reported an elegant stereospecific C-glycosylation method based on sp^3–sp^2 Kosugi–Migita–Stille coupling using anomeric stannane as a donor,[47] and this was later expanded to the synthesis of CF_2-linked disaccharide (Scheme 11-3).[21] The strategy employed in these reports can be regarded as being based on anomeric anion species.

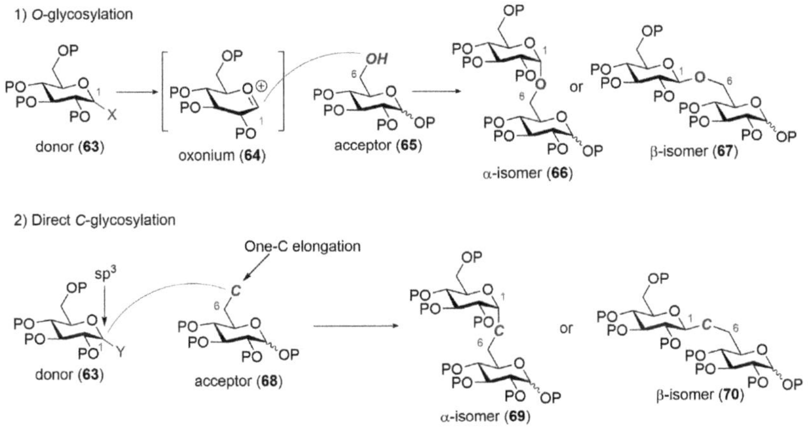

Scheme 12 Schematic representations of (1) O-glycosylation and (2) direct C-glycosylation.

In contrast to those reports, we planned to develop direct C-glycosylation methodologies via anomeric radical species. Anomeric radicals have been used for C-glycosylation for some time,[48] and guidelines for steric control have already been reported (Scheme 13-1).[49] In the case of an anomeric radical having a glucose structure with the usual chair-like conformation ($^4C_1$), orbital interactions (anomeric effect) with the lone pair of oxygen atoms in the ring enhances the reactivity of α–oriented radicals 71, which afford α-C-glycosides. In this case, however, the boat conformation ($B_{2,5}$) is predicted to be more stable, and in addition to the same orbital interactions as before, there may be an interaction with the antibonding orbital of the C-2–O bond (quasi-homo-anomeric effect). In addition to the role of steric factors, C-glycosylation of anomeric radicals arising from glucose structures often shows little stereoselectivity. When the 2-OH group is protected by an acyl group, rearrangement of the acyloxy group to C-1 results in the C-2 radical species with $^4C_1$ conformation, as in **73**. Several examples of stereoselective C-glycosylation have been reported based on these unique properties (Schemes 13-1–4).[50] However, all of them were C-glycosides of monosaccharides or simplified sugar structures, and to our knowledge, no example of disaccharide synthesis based on anomeric radical species had been reported at that time.

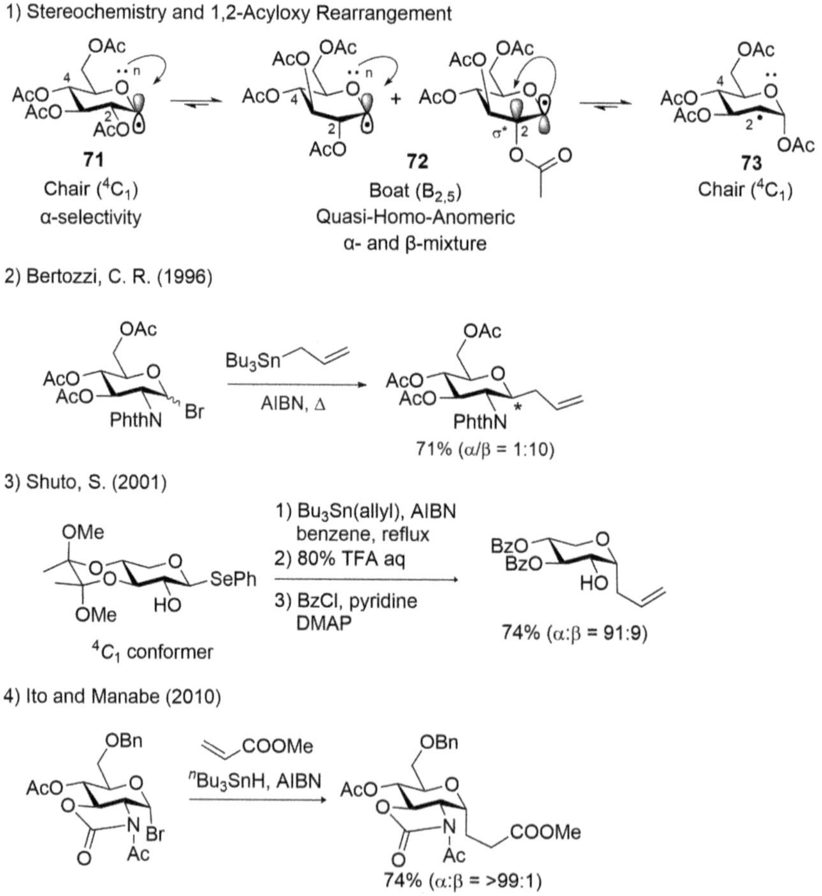
1) Stereochemistry and 1,2-Acyloxy Rearrangement
OAc
AcO
AcO
AcO
71
Chair ($^4C_1$)
α-selectivity
72
Boat ($B_{2,5}$)
Quasi-Homo-Anomeric
α- and β-mixture
73
Chair ($^4C_1$)
2) Bertozzi, C. R. (1996)
Bu_3Sn
AIBN, Δ
PhthN
Br
71% (α/β = 1:10)
3) Shuto, S. (2001)
OMe
SePh
HO
$^4C_1$ conformer
1) Bu_3Sn(allyl), AIBN
benzene, reflux
2) 80% TFA aq
3) BzCl, pyridine
DMAP
BzO
74% (α:β = 91:9)
4) Ito and Manabe (2010)
OBn
COOMe
$^nBu_3SnH$, AIBN
Ac
74% (α:β = >99:1)

Scheme 13 (1) Conformation and 1,2-acyloxy rearrangement of anomeric radicals (in the case of glucose). (2–4) Representative stereoselective C-glycosylation via anomeric radicals.

To confirm the feasibility of using anomeric radicals to direct C-glycosylation, we decided to target simple 1,6-linked disaccharide analogs and investigate atom-transfer radical coupling, as reported by Zard and co-workers (Scheme 14-1).[51] Acetyl-protected glucosyl xanthate donor **74** was treated with lauroyl peroxide in dichloroethane at elevated temperature to afford an anomeric radical **71**. When the radical **71** couples with the acceptor **78**, a new secondary radical **79** would be produced. The reaction of **79** with the donor **74** regenerates the anomeric radical **71** with formation of the desired coupling product **76**. In this system, the relative stability of the anomeric radical **71** compared to **79** should be important, as an imbalance will cause **79** to react with acceptor **78**, not the donor **74**. When the reaction was carried out with acetonide-protected galactose acceptor **75** with a (one-carbon elongated) terminal olefin at the C-6 position, the desired **76** was not formed, and 64% of **77** with a xanthate group at the C-2 position and an acetoxy group at the C-1 position was formed. This compound would be produced by acetoxy group migration of anomeric radical **71** to the C-1 position to form thermodynamically stable C-2 radical **73**, which then reacts with the donor **74**. Therefore, we decided to investigate protecting groups at O-2 other than an acyl group (Scheme 14-2). We found that using **80a** protected at O-2, O-3, and O-4 with TBS groups and acetylated at O-6 allowed the coupling reaction to proceed, giving **81a** in a moderate yield. This indicates that atom-transfer radical coupling can essentially be achieved if the acyloxy rearrangement is suppressed. At this point, the stereoselectivity of the anomeric position could not be determined. After the xanthate group was removed under radical conditions to give **82a**, we could determine the selectivity to be 5:3 (α:β). This result suggested that the resulting anomeric radicals mainly adopt boat conformation, which is considered to lead to an α- and β-isomer mixture. In order to control the conformation to $^4C_1$, a cyclic protecting group was introduced. The donor **80b** with O-4,6-cyclic protection did not work, but **80c** with additional O-2,3-carbonate protection afforded **82c** in 29% yield with high α-selectivity.[1,10] This is the first example of the synthesis of a CH_2-linked disaccharide analog by direct C-glycosylation through anomeric radicals, to our knowledge.

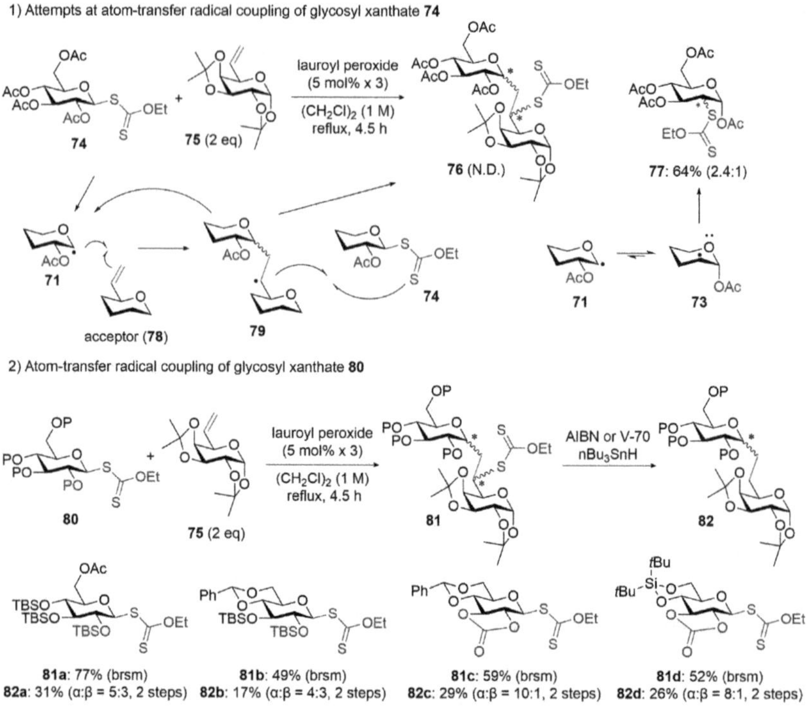

1) Attempts at atom-transfer radical coupling of glycosyl xanthate **74**

2) Atom-transfer radical coupling of glycosyl xanthate **80**

81a: 77% (brsm)   **81b: 49% (brsm)**   **81c: 59% (brsm)**   **81d: 52% (brsm)**
82a: 31% (α:β = 5:3, 2 steps)   **82b: 17% (α:β = 4:3, 2 steps)**   **82c: 29% (α:β = 10:1, 2 steps)**   **82d: 26% (α:β = 8:1, 2 steps)**

Scheme 14 (1) Atom-transfer radical coupling of **74** and a plausible mechanism. (2) Protecting group effects in atom-transfer radical coupling of **80**.

There is a limited choice of protecting groups that can be converted while retaining the xanthate group at the C-1 position. Moreover, since a high reaction concentration (such as 1 M) was essential for this coupling reaction, the solubility of both the donor and the acceptor must also be taken into consideration. When this reaction was applied to the synthesis of the isomaltose (Glc-α-1,6-Glc) analog, the yield of the coupling reaction varied depending on the protecting group of the acceptor. Finally, in order to reduce steric hindrance around the terminal olefin of the acceptor, we employed the acceptor **83** with an acetonide group at O-1 and O-2, and TES groups, which have less electron-withdrawing ability and excellent solubility, at O-3 and O-4. Furthermore, **80d**, which has a di-*tert*-butylsilylene group on O-4 and O-6 of the donor, was found to be the best, and coupling with **83** gave the desired **84** in 69% yield (after **80d** and **83** were recovered following the first reaction and re-coupled again). Reductive removal of the xanthate group gave **85** in a highly α-selective manner in good yield. All protecting groups of **85** were removed in a stepwise manner and the resulting **87** was per-acetylated for purification to give **86**. The acetyl groups were removed by methanolysis to complete the synthesis of the unprotected CH_2-linked isomaltose analog (Scheme 15). Synthesis of a protected CH_2-linked isomaltose analog has been reported, but this is the first example of an unprotected analog.[52] Recently, we have found that pseudo-isomaltose **87** has unique biological activities, supporting the value of C-glycoside analogs. Details will be presented elsewhere.

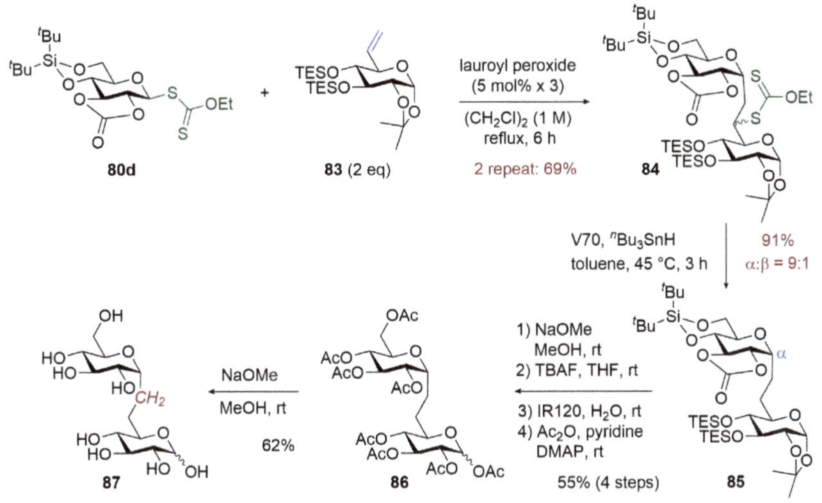

Scheme 15 Synthesis of pseudo-isomaltose (**87**) by atom-transfer radical coupling of **80d** and **83**.

KRN7000 (**88**) is a glycolipid having an α-galactosylceramide structure, which binds to CD1d on antigen-presenting cells and is presented to receptors on NKT cells, thereby activating various immune cells and exhibiting anti-tumor activity (Scheme 16-1).[53] Due to its biological activity and unique mode of action, many analogs of **88** have been developed.[54] For example, Flanck and co-workers reported that the in vivo antitumor effect of CH_2-linked KRN7000 analog **89**, in which the O-glycosidic linkage of **88** is replaced by a simple CH_2 group, is enhanced by about 100-fold compared to **88**.[18,55] Several synthetic methodologies for **89** have been developed.[56] In 2004, Flanck and co-workers reported an efficient synthetic method based on a cross-metathesis reaction (Scheme 16-2).[19] Construction of an α-C-glycosidic bond at C-1 of the galactose ring to form **90**, followed by a cross-metathesis reaction with the terminal olefin of sphingosine analog **91**, provided the CH_2-linked glycolipid structure **92**. Subsequently, a four-step process involving N-acylation led to **89**. Nevertheless, there remains room for improvement in the synthetic strategy, because the construction of the C-glycosidic linkage and the introduction of the sphingosine moiety are done separately. We investigated the possibility of applying our direct C-glycosylation method to the synthesis of CH_2-linked KRN7000 analog **89**. Although the synthesis of analogs of **89** has been reported, the direct C-glycosylation strategy was expected to provide access to a more diverse range of analogs.[57]

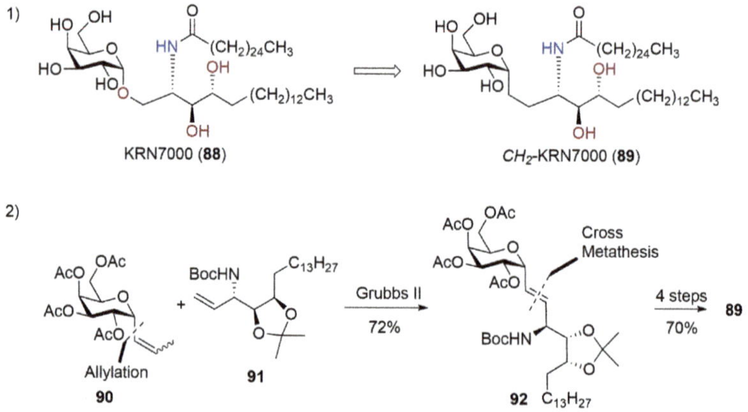

Scheme 16 (1) Structures of KRN7000 (**88**) and CH_2-linked KRN7000 (**89**). (2) Representative synthetic methodology for **89**, as reported by Ref. 19.

Therefore, we investigated the direct C-glycosylation of the $^4C_1$-restricted glucose-type donor **80d** with the acceptor **93** prepared from phytosphingosine (Scheme 17). Since the steric hindrance of the acceptor directly affected the C-glycosylation during the synthesis of the isomaltose analog **87**, we first planned to employ as compact a protecting group as possible. Thus, we prepared **93a** with an acetonide group for the protection of two hydroxyl groups and an N-Boc group and coupled it with **80d**. To our surprise, no coupling product was observed at all. At this point, the reason for this was not clear, and to investigate the effect of the presence of a hydrogen atom on the nitrogen, **93b** was synthesized with a phthaloyl group on the nitrogen atom. In the coupling reaction with **80d**, **94b** was produced in 29% yield (46% allowing for recovered starting material). This result strongly suggested that the presence of NH prevented the coupling reaction, and we anticipated that aggregate formation via intermolecular hydrogen bonding between acceptors **93** might block access to the donor. If this is the case, converting the protective group of the hydroxyl group to a bulkier one could prevent aggregate formation and allow the coupling reaction to proceed more efficiently. The acceptor **93c** with Bz groups was not very effective, but **93d** with TMS groups showed greatly improved coupling efficiency, giving **94d** in 53% yield (89% allowing for recovered starting material). The yield of **93e** with the TBS group was lower, suggesting that the reaction efficiency was reduced if the protecting group was too bulky. Therefore, we employed the acceptor **93f** with TES groups and found that the starting materials were completely consumed, giving **94f** in 81% yield. After reduction of the xanthate group (**95f**; 57% in two steps), we found that α-selectivity at the anomeric center (α:β = 7:2) was lower than expected. Although the reason for this remains unclear, we consider it a great success that we were able to obtain glycolipid structures with adequate stereoselectivity by direct C-glycosylation of such large units.

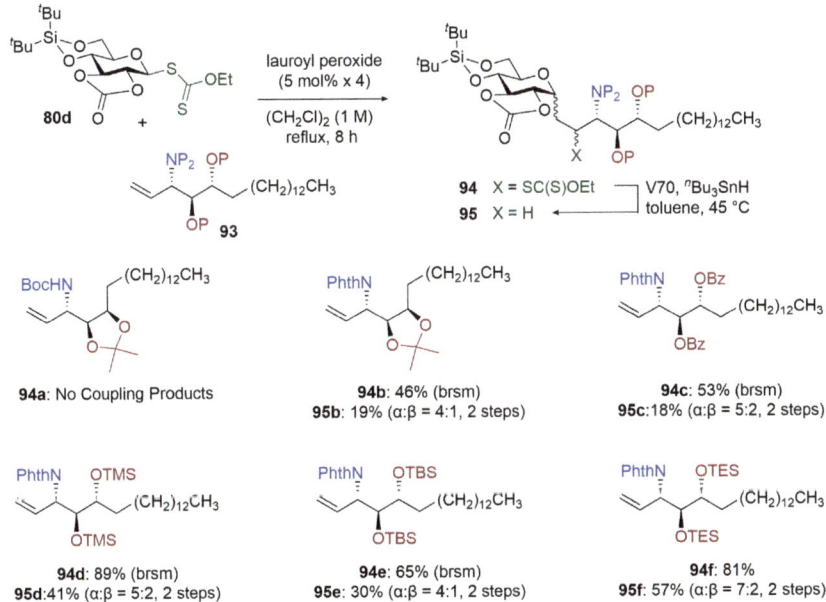

Scheme 17 Direct C-glycosylation of **80d** with sphingosine derivatives **93**.

The synthesis of conformationally restricted galactose donor **96** with cyclic carbonate protection at O-2 and O-3 and subsequent direct C-glycosylation with the optimized sphingosine acceptor **93f** afforded the protected α-galatosylsphingosine derivative **97** in reasonable yield after reduction of the xanthate group (Scheme 17). Removal of the carbonate group followed by the phthaloyl group gave the lyso α-galatosylsphingosine derivative. N-Acylation with the cerotyl chloride and removal of all silyl groups completed the synthesis of CH$_2$-linked KRN7000 (**89**). The low yields of the last four steps were probably due to the moderate yield of the N-acylation step, as well as the loss of a part of desired product during purification of **89** for some reason. Although these processes will need to be improved in the future, this is the first example of the synthesis of CH$_2$-linked KRN7000 (**89**) (Scheme 18) by direct C-glycosylation.[58]

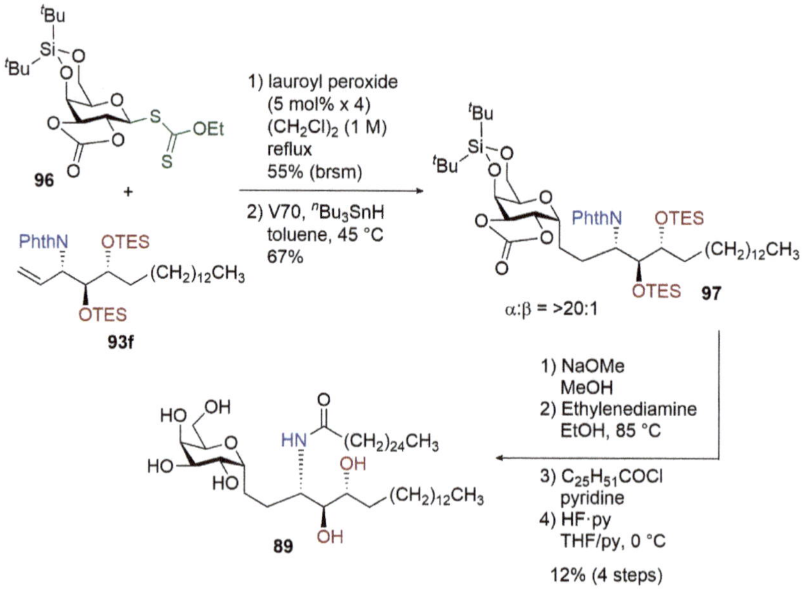

Scheme 18 Synthesis of CH$_2$-linked KRN7000 (**89**) by direct C-glycosylation.

8. Conclusions

This paper summarizes our recent studies aimed at the development of glycoconjugate analogs that are not cleaved by glycoside hydrolases in cells by replacing the potentially cleavable O-glycosidic linkage of the glycoconjugate with a C-glycosidic linkage. In addition, we aimed to make these analogs conformationally controllable by introducing an F atom into the C-glycoside linkage. In the second half of this paper, we summarize the development of a direct C-glycosylation method using atom-transfer radical coupling, and the synthesis of CH$_2$-linked isomaltose and KRN7000 analogs. Although this method is superior to others from the viewpoint of a convergent synthesis, it still has issues in terms of efficiency. Also, it is not applicable to synthesize CHF-linked analogs, and thus new methods are still needed. We have already developed a more efficient synthetic method using different coupling techniques, and a part of this work has already been published. In the future, we aim to further develop this method and utilize it to evaluate the biological activities of pseudo-glycoconjugates with a C-glycoside linkage, as this has so far been achieved only with GM3 analogs.

Finally, the author would like to thank all students, postdoctoral fellows, and collaborators who worked together on this study, especially Prof. Mikiko Sodeoka (RIKEN, Japan). The author would like to express his deepest gratitude to the late Prof. Hidetoshi Yamada, who was always available to give generous advice during our research. This research was supported by AMED, JSPS, RIKEN, Sumitomo Foundation, Terumo Life Science Foundation, and Mizutani Foundation for Glycoscience.

References

1. Varki, A.; Cummings, R. D.; Esko, J.; Stanley, P.; Hart, G. W.; Aebi, M.; Seeberger, P. H., Eds. *Essentials of Glycobiology*; 4th ed.; Cold Spring Harbor Laboratory Press: New York, 2022.
2. Monie, T. P. *Innate Immune System—A Compositional and Functional Perspective*; Elsevier/ Academic Press: London, 2017.
3. Rauter, A. P.; Christensen, B. E.; Somsák, L.; Kosma, P.; Adamo, R. *Recent Trends in Carbohydrate Chemistry: Synthesis and Biomedical Applications of Glycans and Glycoconjugates*; Elsevier: Amsterdam, 2020.
4. Ziegler, S.; Pries, V.; Hedberg, C.; Waldmann, H. Target Identification for Small Bioactive Molecules: Finding the Needle in the Haystack. *Angew. Chem., Int. Ed.* **2013**, *52*, 2744–2792.
5. Miyagi, T.; Takahashi, K.; Yamamoto, K.; Shiozaki, K.; Yamaguchi, K. Biological and Pathological Roles of Ganglioside Sialidases. In *Progress in Molecular Biology and Translational Science*; Schnaar, R. L., Lopez, P. H. H., Eds.; vol. 156; Elsevier/ Academic Press: London, 2018; pp. 121–150. Chapter 4.
6. Geissner, A.; Baumann, L.; Morley, T. J.; Wong, A. K. O.; Sim, L.; Rich, J. R.; So, P. P. L.; Dullaghan, E. M.; Lessard, E.; Iqbal, U.; Moreno, M.; Wakarchuk, W. W.; Withers, S. G. 7-Fluorosialyl Glycosides Are Hydrolysis Resistant but Readily Assembled by Sialyltransferases Providing Easy Access to More Metabolically Stable Glycoproteins. *ACS Cent. Sci.* **2021**, 7, 345–354.
7. (a) Ogawa, S.; Yoshikawa, M.; Taki, T. Synthesis of a Carbocyclic Analog of *N*-Acetylneuraminic Acid (Pseudo-*N*-acetylneuraminic Acid). *J. Chem. Soc., Chem. Commun.* **1992**, 106; (b) Ogawa, S.; Yoshikawa, M.; Taki, T.; Yokoi, S.; Chida, N. Synthesis of Carbocyclic Analogs of 3-Deoxy-d-manno-2-octulosonic Acid and *N*-Acetylneuraminic Acid. *Carbohydr. Res.* **1995**, *269*, 53–78; (c) Mohan, S.; Thompson, J. R.; Pinto, B. M.; Bennet, A. J. Versatile Synthetic Route to Carbocyclic *N*-Acetylneuraminic Acid and Its Derivatives. *Tetrahedron* **2018**, *74*, 5213–5221.

8. (a) Chokhawala, H. A.; Cao, H.; Yu, H.; Chen, X. Enzymatic Synthesis of Fluorinated Mechanistic Probes for Sialidases and Sialyltransferases. *J. Am. Chem. Soc.* **2007**, *129*, 10630–10631; (b) Hayashi, T.; Kehr, G.; Bergander, K.; Gilmour, R. Stereospecific α-Sialylation by Site-Selective Fluorination. *Angew. Chem., Int. Ed.* **2019**, *58*, 3814–3818; (c) Lo, H.-J.; Krasnova, L.; Dey, S.; Cheng, T.; Liu, H.; Tsai, T.-I.; Wu, K. B.; Wu, C.-Y.; Wong, C.-H. Synthesis of Sialidase-Resistant Oligosaccharide and Antibody Glycoform Containing α2,6-Linked 3Fax-Neu5Ac. *J. Am. Chem. Soc.* **2019**, *141*, 6484–6488.

9. Ito, Y.; Kiso, M.; Hasegawa, A. Studies on the Thioglycosides of *N*-Acetylneuraminic Acid 6. Synthesis of Ganglioside GM4 Analogs. *J. Carbohydr. Chem.* **1989**, *8*, 285–294.

10. Ladisch, S.; Hasegawa, A.; Li, R.; Kiso, M. Immunosuppressive Activity of Chemically Synthesized Gangliosides. *Biochemistry* **1995**, *34*, 1197–1202.

11. Rumble, J. R., Jr.; Bruno, T. J.; Doa, M. J. *CRC Handbook of Chemistry and Physics*; CRC Press/Taylor and Francis: Boca Raton, 2021.

12. Hirai, G.; Watanabe, T.; Yamaguchi, K.; Miyagi, T.; Sodeoka, M. Stereocontrolled and Convergent Entry to CF_2-Sialosides: Synthesis of CF_2-Linked Ganglioside GM4. *J. Am. Chem. Soc.* **2007**, *129*, 15420–15421.

13. Poveda, A.; Asensio, J. L.; Polat, T.; Bazin, H.; Linhardt, R.J.; Jiménez-Barbero, J. Conformational Behavior of C-Glycosyl Analogues of Sialyl-α-(2→3)-galactose. *Eur. J. Org. Chem.* **2000**, *2000*, 1805–1813.

14. (a) Ha, Y.; Stevens, D. J.; Skehel, J. J.; Wiley, D. C. X-Ray Structures of H5 Avian and H9 Swine Influenza Virus Hemagglutinins Bound to Avian and Human Receptor Analogs. *Proc. Natl. Acad. Sci. U. S. A.* **2001**, *98*, 11181–11186; (b) Gamblin, S. J.; Haire, L. F.; Russell, R. J.; Stevens, D. J.; Xiao, B.; Ha, Y.; Vasisht, N.; Steinhauer, D. A.; Daniels, R. S.; Elliot, A.; Wiley, D. C.; Skehel, J. J. The Structure and Receptor Binding Properties of the 1918 Influenza Hemagglutinin. *Science* **2004**, *303*, 1838–1842; (c) Lin, T.; Wang, G.; Li, A.; Zhang, Q.; Wu, C.; Zhang, R.; Cai, Q.; Song, W.; Yuen, K.-Y. The Hemagglutinin Structure of an Avian H1N1 Influenza A Virus. *Virology* **2009**, *392*, 73–81.

15. Reiter, D. M.; Frierson, J. M.; Halvorson, E. E.; Kobayashi, T.; Dermody, T. S.; Stehle, T. Crystal Structure of Reovirus Attachment Protein σ1 in Complex With Sialylated Oligosaccharides. *PLoS Pathog.* **2011**, *7*, e1002166.

16. Yang, Y.; Yu, B. Recent Advances in the Chemical Synthesis of C-Glycosides. *Chem. Rev.* **2017**, *117*, 12281–12356.

17. Kitamura, K.; Ando, Y.; Matsumoto, T.; Suzuki, K. Total Synthesis of Aryl C-Glycoside Natural Products: Strategies and Tactics. *Chem. Rev.* **2018**, *118*, 1495–1598.

18. (a) Tomiyama, H.; Yanagisawa, T.; Nimura, M.; Noda, A.; Tomiyama, T. *Preparation of Ceramidyl Glycosides or C-Glycosides for Use as Antitumor Agents or Immunostimulants*; DE10128250A1, 2001; (b) Yang, G.; Schmieg, J.; Tsuji, M.; Franck, R. W. The C-Glycoside Analogue of the Immunostimulant α-Galactosylceramide (KRN7000): Synthesis and Striking Enhancement of Activity. *Angew. Chem., Int. Ed.* **2004**, *43*, 3818–3822. Patent Number is DE10128250A1.

19. Chen, G.; Schmieg, J.; Tsuji, M.; Franck, R. W. Efficient Synthesis of α-C-Galactosyl Ceramide Immunostimulants: Use of Ethylene-Promoted Olefin Cross-Metathesis. *Org. Lett.* **2004**, *6*, 4077–4080.

20. (a) Jimenez-Barbero, J.; Demange, R.; Schenk, K.; Vogel, P. Synthesis and Solution Conformational Analysis of 2,3-Anhydro-3-*C*-[(1*R*)-2,6-anhydro-1-deoxy-1-fluoro-D-*glycero*-D-*gulo*-heptitol-1-*C*-yl]-β-D-*gulo*-furanose: First Example of a Monofluoromethylene-Linked C-Disaccharide. *J. Org. Chem.* **2001**, *66*, 5132–5138; (b) Picard, J.; Lubin-Germain, N.; Uzie, J.; Auge, J. Indium-Mediated Alkynylation in C-Glycoside Synthesis. *Synthesis* **2006**, 979–982; (c) Perez-Castells, J.;

Hernandez-Gay, J. J.; Denton, R. W.; Tony, K. A.; Mootoo, D. R.; Jimenez-Barbero, J. The Conformational Behavior and P-Selectin Inhibition of Fluorine-Containing Sialyl Lex Glycomimetics. *Org. Biomol. Chem.* **2007**, *5*, 1087–1092; (d) Tony, K. A.; Denton, R. W.; Dilhas, A.; Jimenez-Barbero, J.; Mootoo, D. R. Synthesis of β-*C*-Galacto-pyranosides With Fluorine on the Pseudoanomeric Substituent. *Org. Lett.* **2007**, *9*, 1441–1444; (e) Colombel, S.; Sanselme, M.; Leclerc, E.; Quirion, J.-C.; Pannecoucke, X. Straightforward Preparation of Functionalized α-CF$_2$-Galactosides Through an Oxygen to Carbon Acyl Migration. *Chem. - Eur. J.* **2011**, *17*, 5238–5241; (f) Colombel, S.; Van Hijfte, N.; Poisson, T.; Leclerc, E.; Pannecoucke, X. Addition of Electrophilic Radicals to 2-Benzyloxyglycals: Synthesis and Functionalization of Fluorinated α-C-Glycosides and Derivatives. *Chem. - Eur. J.* **2013**, *19*, 12778–12787; (g) Leclerc, E.; Pannecoucke, X.; Etheve-Quelquejeu, M.; Sollogoub, M. Fluoro-C-glycosides and Fluoro-carbasugars, Hydrolytically Stable and Synthetically Challenging Glycomimetics. *Chem. Soc. Rev.* **2013**, *42*, 4270–4283; (h) Altiti, A. S.; Bachan, S.; Alrowhani, W.; Mootoo, D. R. An Organo-Catalytic Strategy for the Stereoselective Synthesis of C-Galactosides With Fluorine at the Pseudo-Anomeric Carbon. *Org. Biomol. Chem.* **2015**, *13*, 10328–10335.

21. Zhu, F.; Rodriguez, J.; O'Neill, S.; Walczak, M. A. Acyl Glycosides Through Stereospecific Glycosyl Cross-Coupling: Rapid Access to C(sp3)-Linked Glycomimetics. *ACS Cent. Sci.* **2018**, *4*, 1652–1662.

22. Werschkun, B.; Thiem, J. Synthesis of Novel Types of Divalent Saccharide Structures by a Ketene Acetal Claisen Rearrangement. *Tetrahedron: Asymmetry* **2005**, *16*, 569–576.

23. van Steenis, J. H.; von der Gen, A. Synthesis of Terminal Monofluoro Olefins. *J. Chem. Soc., Perkin Trans. 1* **2002**, 2117–2133.

24. Hoffmann, R. W. Allylic 1,3-Strain as a Controlling Factor in Stereoselective Transformations. *Chem. Rev.* **1989**, *89*, 1841–1860.

25. Habib, S.; Larnaud, F.; Pfund, E.; Lequeux, T.; Fenet, B.; Goekjian, P. G.; Gueyrard, D. Synthesis of Fluorinated exo-Glycals Through Modified Julia Olefination. *Eur. J. Org. Chem.* **2013**, *2013*, 1872–1875.

26. Lei, X.; Dutheuil, G.; Pannecoucke, X.; Quirion, J.-C. A Facile and Mild Method for the Synthesis of Terminal Bromofluoroolefins via Diethylzinc-Promoted Wittig Reaction. *Org. Lett.* **2004**, *6*, 2101–2104.

27. Watanabe, T.; Hirai, G.; Kato, M.; Hashizume, D.; Miyagi, T.; Sodeoka, M. Synthesis of CH$_2$-Linked α(2,3)Sialylgalactose Analogue: On the Stereoselectivity of the Key Ireland–Claisen Rearrangement. *Org. Lett.* **2008**, *10*, 4167–4170.

28. Sakamoto, H.; Nakamura, S.; Tsuda, T.; Hashimoto, S. Chemoselective Glycosidation Strategy Based on Glycosyl Donors and Acceptors Carrying Phosphorus-Containing Leaving Groups: A Convergent Synthesis of Ganglioside GM3. *Tetrahedron Lett.* **2000**, *41*, 7691–7695.

29. Zeng, Y.; Wang, Z.; Whitfield, D.; Huang, X. Installation of Electron-Donating Protective Groups, a Strategy for Glycosylating Unreactive Thioglycosyl Acceptors Using the Preactivation-Based Glycosylation Method. *J. Org. Chem.* **2008**, *73*, 7952–7962.

30. (a) Yu, B.; Tao, H. Glycosyl Trifluoroacetimidates. Part 1: Preparation and Application as New Glycosyl Donors. *Tetrahedron Lett.* **2001**, *42*, 2405–2407; (b) Yu, B.; Tao, H. Glycosyl Trifluoroacetimidates. 2. Synthesis of Dioscin and Xiebai Saponin I. *J. Org. Chem.* **2002**, *67*, 9099–9102.

31. Crich, D.; Vinogradova, O. Synthesis and Glycosylation of a Series of 6-Mono-, Di-, and Trifluoro S-Phenyl 2,3,4-Tri-O-benzyl-thiorhamnopyranosides. Effect of the Fluorine Substituents on Glycosylation Stereoselectivity. *J. Am. Chem. Soc.* **2007**, *129*, 11756–11765.

32. Li, Y.; Yang, Y.; Yu, B. An Efficient Glycosylation Protocol With Glycosyl *ortho*-Alkynylbenzoates as Donors Under the Catalysis of Ph₃PAuOTf. *Tetrahedron Lett.* **2008**, *49*, 3604–3608.

33. Zhang, Q.; Sun, J.; Zhu, Y.; Zhang, F.; Yu, B. An Efficient Approach to the Synthesis of Nucleosides: Gold(I)-Catalyzed N-Glycosylation of Pyrimidines and Purines With Glycosyl *ortho*-Alkynyl Benzoates. *Angew. Chem., Int. Ed.* **2011**, *50*, 4933–4936.

34. Crich, D. Mechanism of a Chemical Glycosylation Reaction. *Acc. Chem. Res.* **2010**, *43*, 1144–1153.

35. (a) Breg, J.; Kroon-Batenburg, L. M. J.; Strecker, G.; Montreuil, J.; Vliegenthart, J. F. G. Conformational Analysis of the Sialyl(2→3/6)N-acetyllactosamine Structural Element Occurring in Glycoproteins, by Two-Dimensional NOE Proton NMR Spectroscopy in Combination With Energy Calculations by Hard-Sphere exo-Atomic and Molecular Mechanics Force-Field With Hydrogen-Bonding Potential. *Eur. J. Biochem.* **1989**, *178*, 727–739; (b) Mazurier, J.; Dauchez, M.; Vergoten, G.; Montreuil, J.; Spik, G. Molecular Modeling of a Disialylated Monofucosylated Biantennary Glycan of the N-Acetyllactosamine Type. *Glycoconjugate J.* **1991**, *8*, 390–399; (c) Ichikawa, Y.; Lin, Y. C.; Dumas, D. P.; Shen, G. J.; Garcia-Junceda, E.; Williams, M. A.; Bayer, R.; Ketcham, C.; Walker, L. E.; Paulson, J. C.; Wong, C.-H. Chemical-Enzymic Synthesis and Conformational Analysis of Sialyl Lewis X and Derivatives. *J. Am. Chem. Soc.* **1992**, *114*, 9283–9298; (d) Xavier Suresh, M.; Veluraja, K. Conformations of Terminal Sialyloligosaccharide Fragments—A Molecular Dynamics Study. *J. Theor. Biol.* **2003**, *222*, 389–402.

36. Matsumori, N.; Kaneno, D.; Murata, M.; Nakamura, H.; Tachibana, K. Stereochemical Determination of Acyclic Structures Based on Carbon–Proton Spin-Coupling Constants. A Method of Configuration Analysis for Natural Products. *J. Org. Chem.* **1999**, *64*, 866–876.

37. Williamson, R. T.; Marquez, B. L.; Gerwick, W. H.; Martin, G. E.; Krishnamurthy, V. V. J-IMPEACH-MBC: A New Concatenated NMR Experiment for F1 Scalable, J-Resolved HMBC. *Magn. Reson. Chem.* **2001**, *39*, 127–132.

38. Kurz, M.; Schmieder, P.; Kessler, H. HETLOC, an Efficient Method for Determining Heteronuclear Long-Range Couplings With Heteronuclei in Natural Abundance. *Angew. Chem., Int. Ed. Engl.* **1991**, *30*, 1329–1331.

39. Willker, W.; Leibfritz, D. Determination of Heteronuclear Long-Range H,X Coupling Constants From Gradient-Selected HMBC Spectra. *Magn. Reson. Chem.* **1995**, *33*, 632–638.

40. Bifulco, G.; Dambruoso, P.; Gomez-Paloma, L.; Riccio, R. Determination of Relative Configuration in Organic Compounds by NMR Spectroscopy and Computational Methods. *Chem. Rev.* **2007**, *107*, 3744–3779.

41. (a) Li, R.; Manela, J.; Kong, Y.; Ladisch, S. Cellular Gangliosides Promote Growth Factor-Induced Proliferation of Fibroblasts. *J. Biol. Chem.* **2000**, *275*, 34213–34223; (b) Kaucic, K.; Liu, Y.; Ladisch, S. Modulation of Growth Factor Signaling by Gangliosides: Positive or Negative? *Methods Enzymol.* **2006**, *417*, 168–185.

42. Taki, T.; Ogura, K.; Rokukawa, C.; Hara, T.; Kawakita, M.; Endo, T.; Kobata, A.; Handa, S. Had-1, a Uridine 5'-Diphosphogalactose Transport-Defective Mutant of Mouse Mammary Tumor Cell FM3A: Composition of Glycolipids, Cell Growth Inhibition by Lactosylceramide, and Loss of Tumorigenicity. *Cancer Res.* **1991**, *51*, 1701–1707.

43. Hirai, G.; Kato, M.; Koshino, H.; Nishizawa, E.; Oonuma, K.; Ota, E.; Watanabe, T.; Hashizume, D.; Tamura, Y.; Okada, M.; Miyagi, T.; Sodeoka, M. Ganglioside GM3 Analogues Containing Monofluoromethylene-Linked Sialoside: Synthesis, Stereochemical Effects, Conformational Behavior, and Biological Activities. *JACS Au* **2021**, *1*, 137–146.

44. Koester, D. C.; Kriemen, E.; Werz, D. B. Flexible Synthesis of 2-Deoxy-C-glycosides and (1→2)-, (1→3)-, and (1→4)-Linked C-Glycosides. *Angew. Chem., Int. Ed.* **2013**, *52*, 2985–2989.

45. Nielsen, M. M.; Pedersen, C. M. Catalytic Glycosylations in Oligosaccharide Synthesis. *Chem. Rev.* **2018**, *118*, 8285–8358.

46. (a) Miquel, N.; Doisneau, G.; Beau, J.-M. Reductive Samariation of Anomeric 2-Pyridyl Sulfones With Catalytic Nickel: An Unexpected Improvement in the Synthesis of 1,2-*trans*-Diequatorial C-Glycosyl Compounds. *Angew. Chem., Int. Ed.* **2000**, *39*, 4111–4114; (b) Mazéas, D.; Skrydstrup, T.; Doumeix, O.; Beau, J.-M. Samarium Iodide Induced Intramolecular C-Glycoside Formation: Efficient Radical Formation in the Absence of an Additive. *Angew. Chem., Int. Ed. Engl.* **1994**, *33*, 1383–1386; (c) Mazéas, D.; Skrydstrup, T.; Beau, J.-M. A Highly Stereoselective Synthesis of 1,2-*trans*-C-Glycosides via Glycosyl Samarium(III) Compounds. *Angew. Chem., Int. Ed. Engl.* **1995**, *34*, 909–912; (d) Abdallah, Z.; Doisneau, G.; Beau, J.-M. Synthesis of a Carbon-Linked Mimic of the Disaccharide Component of the Tumor-Related Sialyltin Antigen. *Angew. Chem., Int. Ed.* **2003**, *42*, 5209–5212; (e) Malapelle, A.; Abdallah, Z.; Doisneau, G.; Beau, J.-M. Anomeric Acetates of N-Acetylneuraminic Acid Are Useful C-Sialyl Donors in Samarium-Mediated Reformatsky Coupling Reactions. *Angew. Chem., Int. Ed.* **2006**, *45*, 6016–6020; (f) Vlahov, I. R.; Vlahova, P. I.; Linhardt, R. J. Diastereocontrolled Synthesis of Carbon Glycosides of N-Acetylneuraminic Acid via Glycosyl Samarium(III) Intermediates. *J. Am. Chem. Soc.* **1997**, *119*, 1480–1481; (g) Bazin, H. G.; Du, Y.; Polat, T.; Linhardt, R. J. Synthesis of a Versatile Neuraminic Acid "C"-Disaccharide Precursor for the Synthesis of C-Glycoside Analogues of Gangliosides. *J. Org. Chem.* **1999**, *64*, 7254–7259.

47. Zhu, F.; Rourke, M. J.; Yang, T.; Rodriguez, J.; Walczak, M. A. Highly Stereospecific Cross-Coupling Reactions of Anomeric Stannanes for the Synthesis of C-Aryl Glycosides. *J. Am. Chem. Soc.* **2016**, *138*, 12049–12052.

48. Xu, L.-Y.; Fan, N.-L.; Hu, X.-G. Recent Development in the Synthesis of C-Glycosides Involving Glycosyl Radicals. *Org. Biomol. Chem.* **2020**, *18*, 5095–5109.

49. (a) Dupuis, J.; Giese, B.; Rüegge, D.; Fischer, H.; Korth, H.-G.; Sustmann, R. Conformation of Glycosyl Radicals: Radical Stabilization by β-CO Bonds. *Angew. Chem., Int. Ed. Engl.* **1984**, *23*, 896–898; (b) Giese, B.; Witzel, T. Synthesis of "C-Disaccharides" by Radical C–C Bond Formation. *Angew. Chem., Int. Ed. Engl.* **1986**, *25*, 450–451; (c) Korth, H.-G.; Sustmann, R.; Dupuis, J.; Giese, B. Electron Spin Resonance Spectroscopic Investigation of Carbohydrate Radicals. Part 2. Conformation and Configuration in Pyranos-1-yl Radicals. *J. Chem. Soc., Perkin Trans. 2* **1986**, 1453–1459; (d) Giese, B.; Dupuis, J.; Leising, M.; Nix, M.; Lindner, H. J. Synthesis of C-Pento, -Hexo-, and -Heptulo-pyranosyl Compounds via Radical C–C Bond-Formation Reactions. *Carbohydr. Res.* **1987**, *171*, 329–341; (e) Giese, B.; Hoch, M.; Lamberth, C.; Schmidt, R. R. Synthesis of Methylene Bridged C-Disaccharides. *Tetrahedron Lett.* **1988**, *29*, 1375–1378.

50. (a) Roe, B. A.; Boojamra, C. G.; Griggs, J. L.; Bertozzi, C. R. Synthesis of β-C-Glycosides of N-Acetylglucosamine via Keck Allylation Directed by Neighboring Phthalimide Groups. *J. Org. Chem.* **1996**, *61*, 6442–6445; (b) Abe, H.; Shuto, S.; Matsuda, A. Highly α- and β-Selective Radical C-Glycosylation Reactions Using a Controlling Anomeric Effect Based on the Conformational Restriction Strategy A Study on the Conformation−Anomeric Effect−Stereoselectivity Relationship in Anomeric Radical Reactions. *J. Am. Chem. Soc.* **2001**, *123*, 11870–11880; (c) Manabe, S.; Aihara, Y.; Ito, Y. Radical C-Glycosylation Reaction of Pyranosides With the 2,3-*trans* Carbamate Group. *Chem. Commun. (Cambridge, U. K.)* **2011**, *47*, 9720–9722.

51. Zard, S. Z. Radical Alliances: Solutions and Opportunities for Organic Synthesis. *Helv. Chim. Acta* **2019**, *102*, e1900134.

52. Kiya, N.; Hidaka, Y.; Usui, K.; Hirai, G. Synthesis of CH_2-Linked α(1,6)-Disaccharide Analogues by α-Selective Radical Coupling C-Glycosylation. *Org. Lett.* **2019**, *21*, 1588–1592.

53. (a) Kobayashi, E.; Motoki, K.; Uchida, T.; Fukushima, H.; Koezuka, Y. KRN7000, a Novel Immunomodulator, and Its Antitumor Activities. *Oncol. Res.* **1995**, *7*, 529–534; (b) Morita, M.; Motoki, K.; Akimoto, K.; Natori, T.; Sakai, T.; Sawa, E.; Yamaji, K.; Koezuka, Y.; Kobayashi, E.; Fukushima, H. Structure-Activity Relationship of α-Galactosylceramides Against B16-Bearing Mice. *J. Med. Chem.* **1995**, *38*, 2176–2187.

54. (a) Tashiro, T.; Mori, K. Fifteen Years Since the Development of KRN700— Structure–Activity Relationship Studies on Novel Glycosphingolipids Which Stimulate Natural Killer T Cells. *Trends Glycosci. Glycotechnol.* **2010**, *22*, 280–295; (b) Pifferi, C.; Fuentes, R.; Fernández-Tejada, A. Author Correction: Natural and Synthetic Carbohydrate-Based Vaccine Adjuvants and Their Mechanisms of Action. *Nat. Rev. Chem.* **2021**, *5*, 197–216.

55. Schmieg, J.; Yang, G.; Franck, R. W.; Tsuji, M. Superior Protection Against Malaria and Melanoma Metastases by a C-Glycoside Analogue of the Natural Killer T Cell Ligand α-Galactosylceramide. *J. Exp. Med.* **2003**, *198*, 1631–1641.

56. (a) Toba, T.; Murata, K.; Yamamura, T.; Miyake, S.; Annoura, H. A Concise Synthesis of (3S,4S,5R)-1-(α-D-Galactopyranosyl)-3-tetracosanoylamino-4,5-decanediol, a C-Glycoside Analogue of Immunomodulating α-Galactosylceramide OCH. *Tetrahedron Lett.* **2005**, *46*, 5043–5047; (b) Lu, X.; Song, L.; Metelitsa, L. S.; Bittman, R. Synthesis and Evaluation of an α-C-Galactosylceramide Analogue That Induces Th1-Biased Responses in Human Natural Killer T Cells. *ChemBioChem* **2006**, *7*, 1750–1756; (c) Wipf, P.; Pierce, J. G. Expedient Synthesis of the α-C-Glycoside Analogue of the Immunostimulant Galactosylceramide (KRN7000). *Org. Lett.* **2006**, *8*, 3375–3378; (d) Liu, Z.; Byun, H.-S.; Bittman, R. Synthesis of Immunostimulatory α-C-Galactosylceramide Glycolipids via Sonogashira Coupling, Asymmetric Epoxidation, and Trichloroacetimidate-Mediated Epoxide Opening. *Org. Lett.* **2010**, *12*, 2974–2977; (e) Altiti, A. S.; Mootoo, D. R. Intramolecular Nitrogen Delivery for the Synthesis of C-Glycosphingolipids. Application to the C-Glycoside of the Immunostimulant KRN7000. *Org. Lett.* **2014**, *16*, 1472–1475; (f) Altiti, A. S.; Bachan, S.; Mootoo, D. R. The Crotylation Way to Glycosphingolipids: Synthesis of Analogues of KRN7000. *Org. Lett.* **2016**, *18*, 4654–4657; (g) Chang, Y.-J.; Hsuan, Y.-C.; Lai, A. C.-Y.; Han, Y.-C.; Hou, D.-R. Synthesis of α-C-Galactosylceramide via Diastereoselective Aziridination: The New Immunostimulant 4′-*epi*-C-Glycoside of KRN7000. *Org. Lett.* **2016**, *18*, 808–811.

57. (a) Chen, G.; Chien, M.; Tsuji, M.; Franck, R. W. E and Z α-C-Galactosylceramides by Julia–Lythgoe–Kocienski Chemistry: A Test of the Receptor-Binding Model for Glycolipid Immunostimulants. *ChemBioChem* **2006**, *7*, 1017–1022; (b) Tashiro, T.; Hongo, N.; Nakagawa, R.; Seino, K.-I.; Watarai, H.; Ishii, Y.; Taniguchi, M.; Mori, K. RCAI-17, 22, 24–26, 29, 31, 34–36, 38–40, and 88, the Analogs of KRN7000 With a Sulfonamide Linkage: Their Synthesis and Bioactivity for Mouse Natural Killer T Cells to Produce Th2-Biased Cytokines. *Bioorg. Med. Chem.* **2008**, *16*, 8896–8906; (c) Pu, J.; Franck, R. W. C-Galactosylceramide Diastereomers via Sharpless Asymmetric Epoxidation Chemistry. *Tetrahedron* **2008**, *64*, 8618–8629; (d) Li, X.; Chen, G.; Garcia-Navarro, R.; Franck, R. W.; Tsuji, M. Identification of C-Glycoside Analogues That Display a Potent Biological Activity Against Murine and Human Invariant Natural Killer T Cells. *Immunology* **2009**, *127*, 216–225; (e) Liu, Z.; Byun, H.-S.; Bittman, R. Total Synthesis of α-1C-Galactosylceramide, an Immunostimulatory C-Glycosphingolipid, and Confirmation of the Stereochemistry in

the First-Generation Synthesis. *J. Org. Chem.* **2011**, *76*, 8588–8598; (f) Liu, Z.; Courtney, A. N.; Metelitsa, L. S.; Bittman, R. C-Glycosphingolipids With an *exo*-Methylene Substituent: Stereocontrolled Synthesis and Immunostimulation of Mouse and Human Natural Killer T Lymphocytes. *ChemBioChem* **2012**, *13*, 1733–1737; (g) Colombel, S.; Van Hijfte, N.; Poisson, T.; Pannecoucke, X.; Monneaux, F.; Leclerc, E. Synthesis and Immunological Evaluation of Fluorinated α-*C*-Galactosylceramide Analogs. *J. Fluorine Chem.* **2015**, *173*, 84–91; (h) Altiti, A. S.; Ma, X.; Zhang, L.; Ban, Y.; Franck, R. W.; Mootoo, D. R. Synthesis and Biological Activities of C-Glycosides of KRN 7000 With Novel Ceramide Residues. *Carbohydr. Res.* **2017**, *443–444*, 73–77; (i) Guillaume, J.; Seki, T.; Decruy, T.; Venken, K.; Elewaut, D.; Tsuji, M.; Van Calenbergh, S. Synthesis of C6″-Modified α-C-GalCer Analogs as Mouse and Human iNKT Cell Agonists. *Org. Biomol. Chem.* **2017**, *15*, 2217–2225; (j) Ban, Y.; Dong, W.; Zhang, L.; Zhou, T.; Altiti, A. S.; Ali, K.; Mootoo, D. R.; Blaho, V. A.; Hla, T.; Ren, Y.; Ma, X. Abrogation of Endogenous Glycolipid Antigen Presentation on Myelin-Laden Macrophages by D-Sphingosine Ameliorates the Pathogenesis of Experimental Autoimmune Encephalomyelitis. *Front. Immunol.* **2019**, *10*, 404. https://doi.org/10.3389/fimmu.2019.00404.

58. Hidaka, Y.; Kiya, N.; Yoritate, M.; Usui, K.; Hirai, G. Synthesis of CH₂-Linked α-Galactosylceramide and Its Glucose Analogues Through Glycosyl Radical-Mediated Direct C-Glycosylation. *Chem. Commun. (Cambridge, U. K.)* **2020**, *56*, 4712–4715.

CHAPTER FOUR

Boron-mediated aglycon delivery (BMAD) for the stereoselective synthesis of 1,2-*cis* glycosides [☆]

Daisuke Takahashi* and Kazunobu Toshima

Department of Applied Chemistry, Faculty of Science and Technology, Keio University, Yokohama, Japan
*Corresponding author: e-mail address: dtak@applc.keio.ac.jp

Contents

Abbreviations

Ac   acetyl
Ar   aryl
BMAD   boron-mediated aglycon delivery
Bn   benzyl
Boc   *tert*-butoxycarbonyl

[☆]Dedicated to the memory of Prof. Hidetoshi Yoshida.

Advances in Carbohydrate Chemistry and Biochemistry, Volume 82
ISSN 0065-2318
https://doi.org/10.1016/bs.accb.2022.10.003

79

BRSM	based on recovered starting material
Bu	butyl
Bz	benzoyl
CAN	cerium(IV) ammonium nitrate
Cbz	benzyloxycarbonyl
CD	cyclodextrin
Cp	cyclopentadienyl
DCE	1,2-dichloroethane
DFT	density functional theory (calculation)
DMAP	4-(dimethylamino)pyridine
DMF	N,N-dimethylformamide
DMP	2,6-dimethylphenyl
DMSO	dimethyl sulfoxide
DTBS	di-*tert*-butylsilylene
EDB	1,1′-(ethane-1,2-diyl)dibenzene-2,2′-bis(methylene)
EDCI	1-ethyl-3-(3-dimethylaminopropyl)carbodiimide
GSL	glycosphingolipid
HAD	hydrogen bond-mediated aglycon delivery
HFIP	hexafluoroisopropanol
HOBt	1-hydroxybenzotriazole
IAD	intramolecular aglycon delivery
IRC	intrinsic reaction coordinate (scan)
KIE	kinetic isotope effect
LPS	lipopolysaccharide
Me	methyl
MEL	mannosylerythritol lipid
MS	molecular sieves
NIS	N-iodosuccinimide
NMR	nuclear magnetic resonance
Phth	phthaloyl
PMB	p-methoxybenzyl
Py	pyridine
S_N1	unimolecular nucleophilic substitution
S_N2	bimolecular nucleophilic substitution
S_Ni	intramolecular nucleophilic substitution
TBAF	tetra-n-butylammonium fluoride
TBAI	tetra-n-butylammonium iodide
TBDPS	*tert*-butyldiphenylsilyl
TBS	*tert*-butyldimethylsilyl
TESH	triethylsilane
Tf	triflyl (trifluoromethanesulfonyl)
TFA	trifluoroacetic acid
THF	tetrahydrofuran (oxolane)
TMPO	2,2,6,6-tetramethylpipedinyloxy
Tr	2,2,2-trichloroethoxycarbonyl
Troc	2,2,2-trichloroethoxycarbonyl
TS	transition state

1. Introduction

Various glycosides with 1,2-*cis*-glycosidic linkages, such as α-D-glucosides and β-D-mannosides, are found in naturally occurring biologically active compounds, in pharmaceutical compounds, and in highly functional materials (Fig. 1).[1–6] Therefore, elucidating the mechanism of their biological activities will help clarify the structure–activity relationships of these diverse compounds and create new lead compounds for pharmaceuticals by modifying their structures. However, unlike 1,2-*trans* glycosides such as β-D-glucosides and α-D-mannosides, the stereoselective synthesis

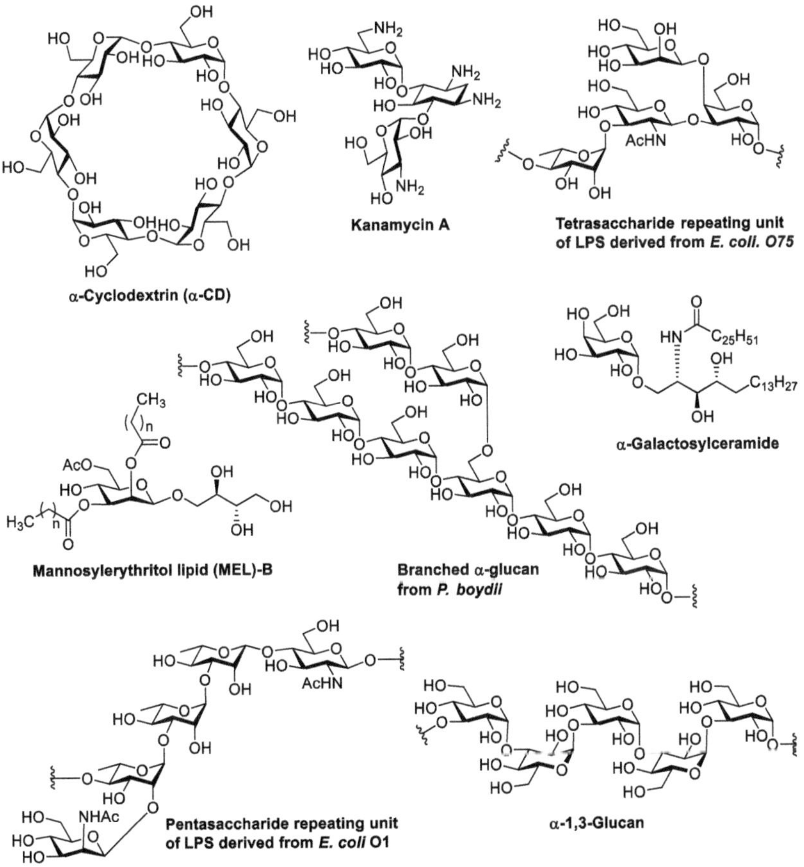

Fig. 1 Chemical structures of several carbohydrates possessing 1,2-*cis* glycosides.

of 1,2-*cis* glycosides remains difficult because of the absence of neighboring group participation from 2-*O*-acyl functionalities of the glycosyl donors. It is therefore important to develop efficient glycosylation methods for the stereoselective synthesis of 1,2-*cis*-glycosides, and several efficient indirect and direct glycosylation methods have been reported to date.[7–32] Representative indirect methods include intramolecular aglycon delivery (IAD),[13,15,16,33–40] a molecular clamp method,[41–43] a prearranged method,[44–54] a peptide-templated method,[55–58] and other intramolecular glycosylation methods.[8,21,59] These methods generally show good to excellent 1,2-*cis*-stereoselectivity due to the intermediate linking of a glycosyl donor and acceptor with a stable tether molecule, but all require additional synthetic steps to prepare the intermediates. In contrast, representative direct methods include hydrogen bond-mediated aglycon delivery (HAD)[31] and conformationally-restrained glycosyl donor-based approaches.[30] For example, Yamada and coworkers reported an elegant 1,2-*cis*-α-stereoselective glycosylation method using a purpose-designed 3,6-*O*-EDB [1,1′-(ethane-1,2-diyl)dibenzene-2,2′-bis(methylene)]-bridged glycosyl donor and its application to the first chemical synthesis of small-ring cyclodextrins CD3 and CD4 (Fig. 2).[60] However, direct methods often do not provide complete 1,2-*cis*-stereoselectivity, and thus the development of a direct and complete 1,2-*cis*-stereoselective glycosylation method is highly desirable. In this context, we recently developed organoboron-catalyzed 1,2-*cis*-stereoselecitve glycosylations, called the boron-mediated

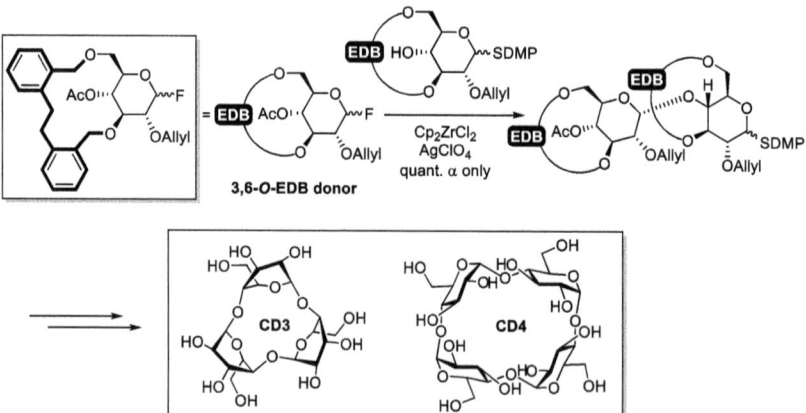

Fig. 2 1,2-*cis*-α-Stereoselective glycosylation using a 3,6-*O*-EDB donor and its application to the synthesis of CD3 and CD4.

aglycon delivery (BMAD) methods.[22,61–68] In this review article, we describe the BMAD methods and several examples of their application to the synthesis of biologically active glycosides.

2. Boron-mediated aglycon delivery (BMAD) using borinic acid catalysts

2.1 Borinic-acid-catalyzed 1,2-*cis*-stereoselective glycosylations

The concept behind stereoselective 1,2-*cis* glycosylation using arylborinic acid catalysts is shown in Fig. 3. We hypothesized that first, arylborinic acid **1** reversibly binds to a mono-ol acceptor **2** to form glycosyl-acceptor-derived borinic ester **3**. Next, borinic ester **3**, which shows sufficient Lewis acidity, activates the 1,2-anhydro donor **4** without any further additives under mild conditions. The oxonium cation-like species thus formed, which involves a tetracoordinate borinate ester **6**, increases the nucleophilicity of the boron-bound oxygen atom,[69,70] and concomitant intramolecular rearrangement provides the 1,2-*cis* glycoside **8** with complete stereoselectivity. A mono-ol exchange reaction between borinic ester **7** and mono-ol acceptor **2** regenerates borinic ester **3** as a catalyst. Therefore, this reaction represents a 1,2-*cis*-stereoselective and catalytic glycosylation method.

To investigate our hypothesis, glycosylations of 1,2-anhydro-3,4,6-tri-O-benzyl-D-glucose (**9**)[71] and 6-O-benzylhexan-1,6-diol (a benzyl-protected "mono-ol," **10**) using catalytic amounts of arylborinic acids were examined under various reaction conditions. The results showed that

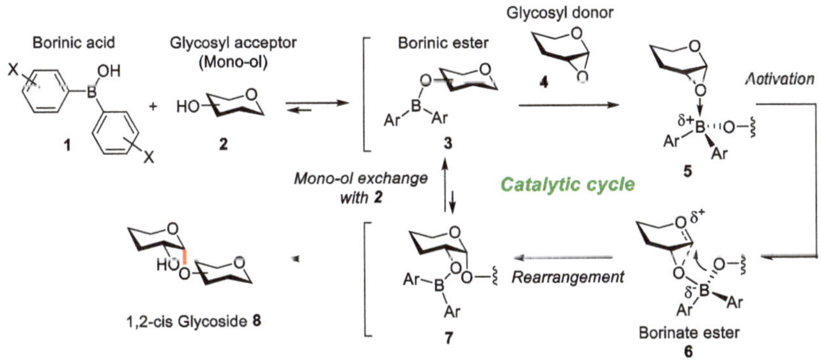

Fig. 3 Concept for a BMAD method using an organoborinic acid.

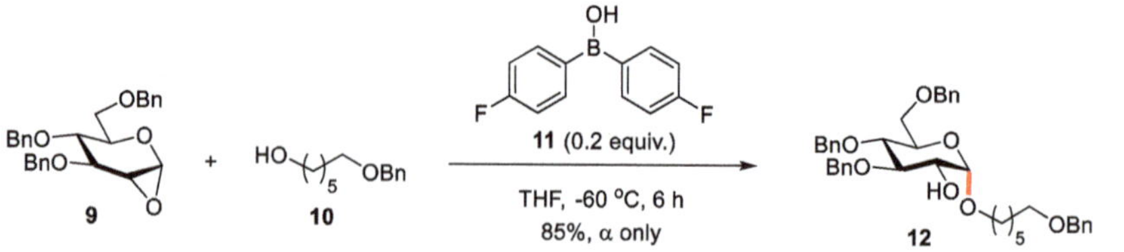

Scheme 1 Stereoselective 1,2-*cis*-α-glycosylation of **9** and **10** using the BMAD method.

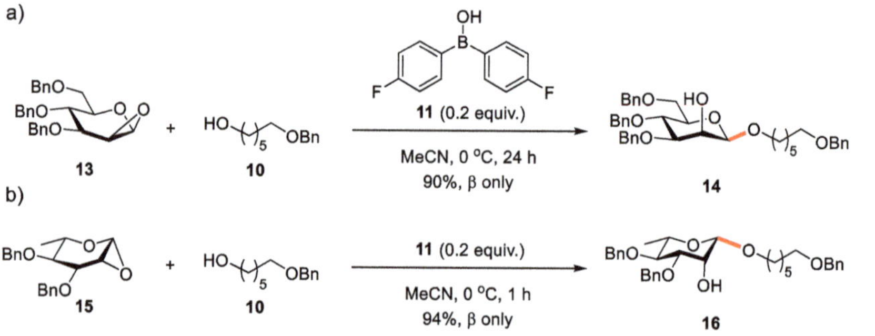

a)

b)

Scheme 2 Stereoselective 1,2-*cis*-β-glycosylations using the BMAD method.

glycosylation of **9** and **10** using di(4–fluoro)phenylborinic acid (**11**)[72] (0.2 equiv.) in THF at −60 °C for 6 h provided the best outcome, producing **12** in 85% yield as a single isomer (Scheme 1).[73]

Next, we applied this BMAD method to the synthesis of more challenging 1,2-*cis* glycoside targets, β-D-mannosides and β-L-rhamnosides. These target compounds are representative of 1,2-*cis*-β-glycosides, since the stereoselective synthesis of 1,2-*cis*-β-glycosides remains challenging due to the nonavailability of neighboring group participation, as well as to the unfavorable anomeric effect and steric hindrance of the axial substituent at the C-2 position. Thus, we first examined glycosylations of 1,2-anhydro-3,4,6-tri-O-benzyl-D-mannose (**13**)[74] and mono-ol **10** using borinic acid catalyst **11**. We found that using MeCN as the reaction solvent, a reaction temperature of 0 °C, and a reaction time of 24 h gave the highest yield of **14** with complete β-stereoselectivity (Scheme 2A).[75] Similarly, we examined glycosylations of 1,2-anhydro-3,4-di-O-benzyl-β-L-rhamnose (**15**)[76] and mono-ol **10** using borinic acid catalyst **11**. After several attempts using various conditions, glycosylation in MeCN at 0 °C for 1 h gave β-L-rhamnoside **16** in high yield as a single isomer (Scheme 2B).[63]

To better understand the glycosylation mechanism, we measured the ^{13}C-KIE (kinetic isotope effect) at the anomeric carbon of the glycosyl donor using quantitative ^{13}C NMR techniques.[77] The KIE value was 1.0034 (51). However, the KIE values for S_N1, S_Ni-type, and dissociative S_N2 reactions are around 1.00, 1.00, and 1.03, respectively,[61,78–80] indicating that this glycosylation can be regarded as either an S_Ni-type reaction or an S_N1 reaction with a short-lived intermediate. To analyze the reaction mechanism in more detail, we conducted DFT calculations using 1,2-anhydro-3,4-di-O-methyl-β-L-rhamnose (**17**) and MeOH as an acceptor to simplify the chemical structure. The transition state (TS) was found at C-1–O-2 = 2.18 Å and C-1–OMe = 2.64 Å. The calculated ^{13}C-KIE value (1.004) based on this TS structure using QUIVER[81] was close to the experimental value. Moreover, an intrinsic reaction coordinate (IRC) scan of the TS structure and potential energy surfaces (Fig. 4)[63] indicated that this glycosylation reaction proceeds without involving the oxocarbenium cation intermediate, and that the glycosylation mechanism is a concerted S_Ni mechanism.

The scope of this reaction was investigated using several donors (**9**, **13**, and **15**) and acceptors (**18–24**). In all cases, the reactions proceeded smoothly to provide the corresponding glycosides (**25–41**) in high yield with complete stereoselectivity (Fig. 5).[63,73,75]

2.2 Total synthesis of GSL-1

GSL-1 (**50**) was isolated from *Flavobacterium devorans* ATCC 1082919[82] and *Sphingomonas paucimobilis*.[83] GSL-1 induces a significant immune response.

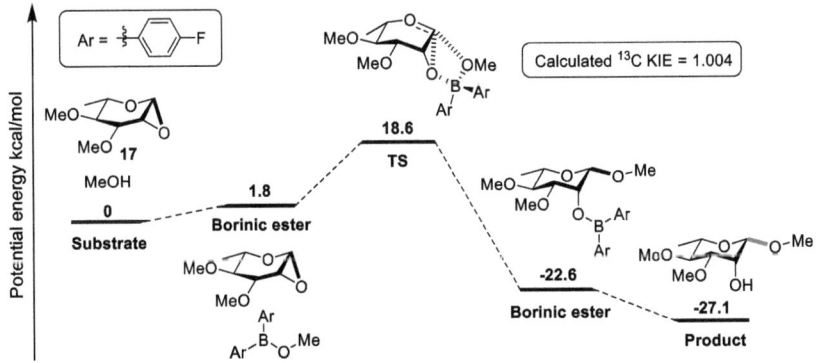

Fig. 4 Potential energy surfaces for the BMAD reaction.

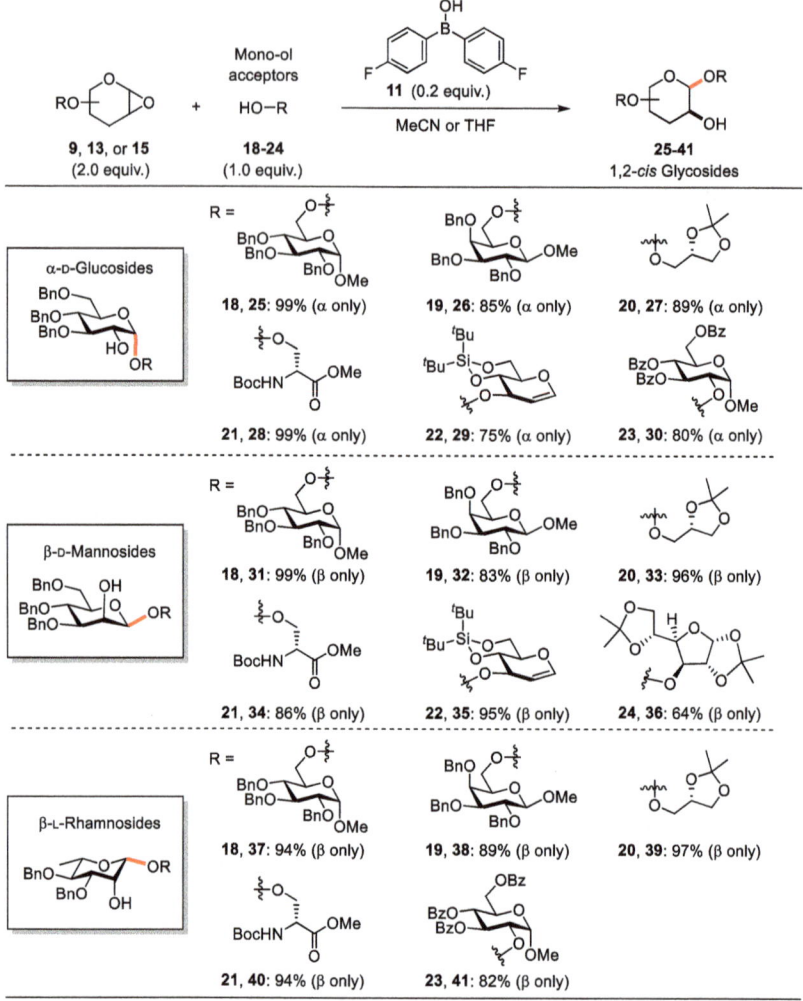

Fig. 5 Glycosylations of several acceptors with **9**, **13**, or **15** using borinic acid **11**.

Structurally, **50** possesses a ceramide moiety and a 1,2-*cis*-α-D-glucuronic acid moiety. Initially, protected ceramide mono-ol acceptor **47** was synthesized from 1-tetradecene (**42**). Known alcohol **43**[84] was prepared from **42** according to the literature. Protection of a hydroxyl group in **43** with a *p*-methoxybenzyl (PMB) group, followed by deprotection of a Tr group, provided **44** in 77% yield in two steps. Oxidation of the resulting primary hydroxyl group in **44**, followed by amidation with known amine **46**,[85]

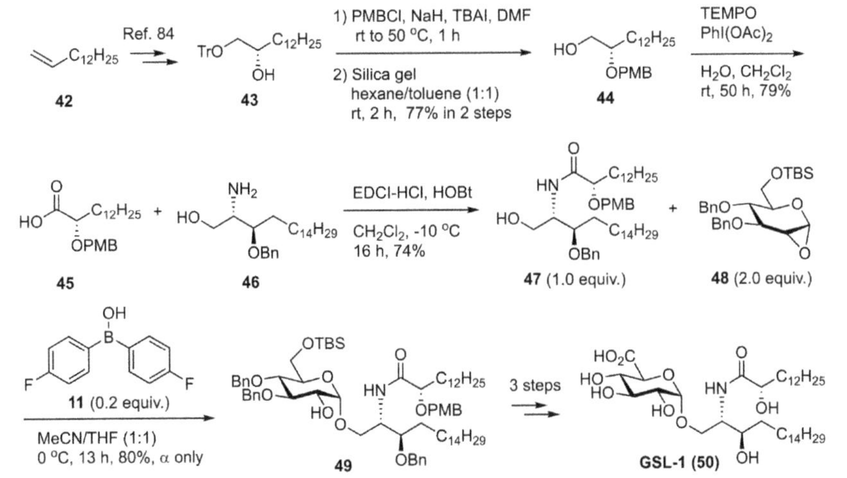

Scheme 3 Total synthesis of GSL-1.

provided the mono-ol **47** in 74% yield. Next, glycosylation of **47** and **48** was examined in the presence of a catalytic amount of **11**. The reaction proceeded smoothly to provide α-glucoside **49** in high yield with complete α-stereoselectivity. Finally, deprotection of the TBS group in **49** and oxidation of the resulting primary hydroxyl group, followed by removal of all protecting groups in three steps, furnished GSL-1 (**50**) (Scheme 3).[73]

2.3 Total synthesis of acremomannolipin A

Acremomannolipin A (**55**)[86] was isolated from *Acremonium strictum* as a potential Ca^{2+} signaling modulator. Structurally, **55** possesses a β-D-mannopyranosyl-D-mannitol moiety and four fatty acyl chains. Initially, protected mannitol mono-ol acceptor **53** was synthesized from D-mannitol (**51**). Known alcohol **52**[87] was prepared from **51** according to the literature procedure. Protection of a hydroxyl group in **52** with a TBS group, followed by deprotection of a Bn group, provided **53** in high yield. Next, **53** and **13** were glycosylated using a catalytic amount of **11** in MeCN at 0 °C for 6 h, and the desired β-D-mannoside **54** was obtained in 99% yield as a single isomer. Finally, the total synthesis of acremomannolipin A (**55**) was accomplished in four steps (Scheme 4).[75]

2.4 Total synthesis of mannosylerythritol lipids (MELs)

Mannosylerythritol lipids (MELs) are natural glycolipid biosurfactants, which are produced by yeast strains such as *Pseudozyma antarctica*.[88]

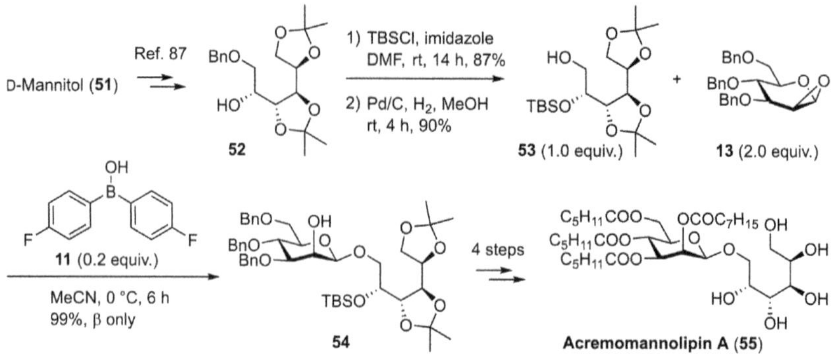

Scheme 4 Total synthesis of acremomannolipin A.

Structurally, MELs possess a β-D-mannopyranosyl-D-erythritol moiety and two fatty acyl chains. In addition, MEL-A, -B, and -C possess one or two acetyl (Ac) groups at C-4′ and/or C-6′ in the mannose moiety, whereas MEL-D lacks Ac groups. The total synthesis of 20 homogeneous members of MEL-A, -B, -C, and -D with different alkyl chain lengths was examined. Initially, protected erythritol acceptor **59** was synthesized from D-glucose (**56**). Known alcohol **57**[89] was prepared from **56** according to the literature procedure. Selective reduction of the benzylidene acetal in **57**, protection of the 1,2-diol with an acetonide group, followed by deprotection of the Bn group, gave **59**. Next, glycosylation of 1,2-anhydro donor **60** and **59** was examined using a catalytic amount of **11** in MeCN at 0 °C for 2 h, producing the desired β-D-mannoside **61** in 99% yield with excellent stereoselectivity. Deprotection of the PMB group in **61** provided the common key intermediate **62**. Acylation of **62** with different length acyl chlorides (C$_6$, C$_8$, C$_{10}$, C$_{12}$ and C$_{14}$), followed by deprotection of the Bn group, gave diols **63–67**, respectively. Acetylation of diols in **63–67**, followed by deprotection of the TBDPS and acetonide groups, furnished MEL-A **68–72**, respectively. Selective acetylation of the primary OH group in **63–67**, followed by deprotection of the TBDPS and acetonide groups, furnished MEL-B **73–77**, respectively. Selective protection of the primary OH group in **63–67** with a TBS group, followed by acetylation and deprotection of the TBDPS and acetonide groups, furnished MEL-C **78–82**, respectively. Finally, deprotection of the TBDPS and acetonide groups of **63–67** furnished MEL-D **83–87**, respectively (Scheme 5).[62]

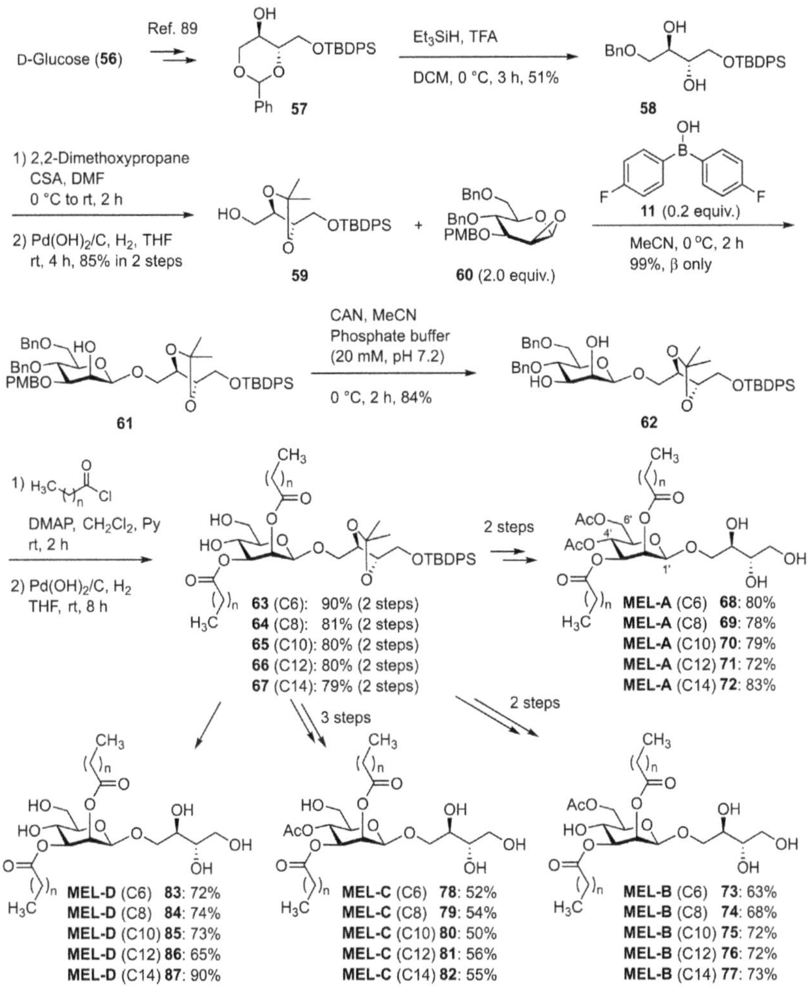

Scheme 5 Total synthesis of mannosylerythritol lipids (MELs).

3. Boron-mediated aglycon delivery (BMAD) using boronic acid catalysts

3.1 Regio- and 1,2-*cis*-stereoselective glycosylations

Next, we focused on regioselective glycosylation and developed a BMAD method using an arylboronic acid, which can bind to a diol acceptor. The

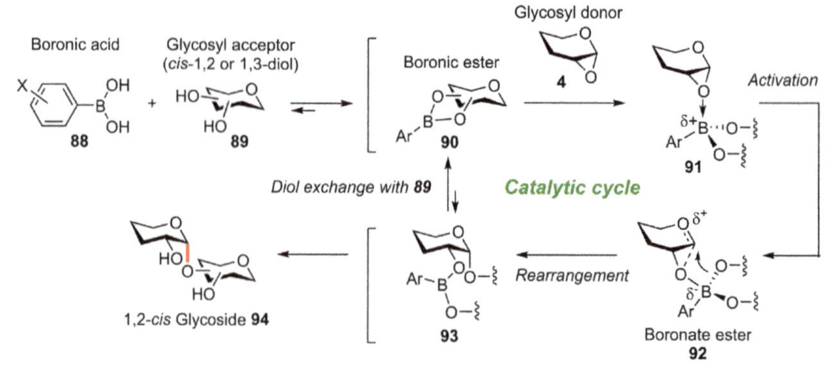

Fig. 6 Concept for BMAD using an organoboronic acid.

concept behind regio- and 1,2-*cis*-stereoselective glycosylation using arylboronic acid catalysts is shown in Fig. 6. We hypothesized that, first, arylboronic acid **88** reversibly binds to a *cis*-1,2- or 1,3-diol glycosyl acceptor **89** to give the glycosyl acceptor-derived-boronic ester **90**. Next, boronic ester **90** activates the 1,2-anhydro glycosyl donor **4** in a manner similar to the borinic-ester-catalyzed method. The formed tetracoordinate boronate ester **92** increases the nucleophilicity of the boron-bound oxygen atom, and concomitant intramolecular glycosylation from the less-hindered oxygen atom provides 1,2-*cis* glycoside **94** with high regio- and 1,2-*cis*-stereoselectivity. Furthermore, a diol exchange reaction between boronic ester intermediate **93** and diol acceptor **89** regenerates boronic ester **90** as a catalyst. Thus, this glycosylation reaction is a regioselective, 1,2-*cis*-stereoselective, catalytic glycosylation method.

We investigated our hypothesis by examining the glycosylation of 1,2-anhydroglucose **9** and several diol acceptors using catalytic amounts of arylboronic acids. For example, glycosylation of **9** and 4,6-diol acceptor **95** using boronic ester catalyst **97**, which was prepared from **95** and **96** under toluene reflux conditions, in MeCN at room temperature for 24 h proceeded readily to give α(1,4)-glycoside **98** in 82% yield with excellent regio- and α-stereoselectivity (Scheme 6A).[90] Next, this BMAD method was applied to regio- and 1,2-*cis*-β-stereoselective glycosylations. The glycosylation of 1,2-anhydromannose **13** and 4,6-diol **99** in the presence of boronic acid catalyst **100** in MeCN at 0 °C for 12 h proceeded regio- and β-stereoselectively to provide β(1,4)-glycoside **102** in 86% yield as a single isomer (Scheme 6B).[91] Similarly, we found that glycosylation of 1,2-anhydro-L-rhamnose **15** and 4,6-diol **103** in the presence of boronic acid catalyst **100**

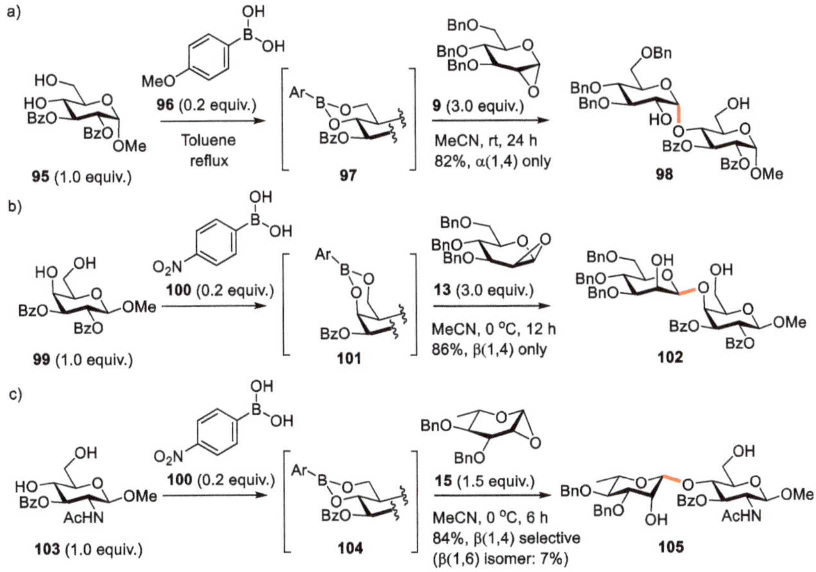

Scheme 6 Regio- and 1,2-*cis*-stereoselective glycosylations using BMAD.

in MeCN at 0 °C for 6 h proceeded efficiently to give β(1,4)-L-glycoside **105** in 84% yield with high regio- and complete β-stereoselectivity (Scheme 6C).[63]

3.2 Synthesis of a tetrasaccharide repeating unit of LPS derived from *Escherichia coli* O75

E. coli O75 causes urinary tract infection and related diseases,[92] and multidrug-resistant strains are emerging.[93,94] Thus, the development of an effective vaccine against *E. coli* O75 is desirable. In this context, we focused on a tetrasaccharide repeating unit of lipopolysaccharide (LPS) derived from *E. coli* O75[95,96] as a glycotope candidate and examined the synthesis of octyl tetrasaccharide glycoside **113** using the BMAD method. Structurally, **113** possesses β(1,4)-D-mannosidic and α-D-galactosidic linkages, which are usually challenging to construct with complete stereoselectivity using a direct method. The glycosylation of 1,2-anhydrogalactose **106**[97] and octanol using borinic acid catalyst **11** (0.2 equiv.) proceeded smoothly to provide galactoside **107** in 74% yield with high α-stereoselectivity. Benzoylation of the 2-OH group in **107**, followed by deprotection of the Bn groups, gave triol **108**. Next, transient masking of the 4,6-diol in **108** using boronic acid **96** under acetone reflux conditions, followed by

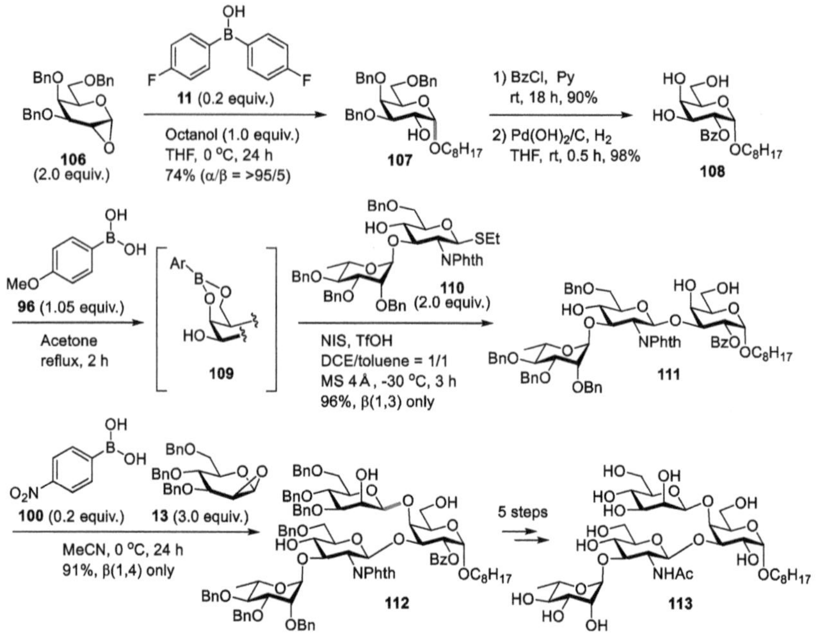

Scheme 7 Synthesis of a tetrasaccharide repeating unit of LPS derived from *E. coli* O75.

glycosylation with disaccharide donor **110** using MS 4 Å in DCE/toluene at −30 °C for 3 h, provided trisaccharide **111** in 96% yield with excellent regio- and stereoselectivity. Finally, conversion of the *N*-Phth group into an *N*-Ac group, followed by removal of the protecting groups, in a total of five steps furnished **113** (Scheme 7).[91]

3.3 Synthesis of a pentasaccharide repeating unit of LPS derived from virulent *E. coli* O1

Next, we demonstrated the first synthesis of **123**, which contains a pentasaccharide repeating unit of LPS derived from virulent *E. coli* O1[98] utilizing the BMAD method. Structurally, **123** contains a β(1,4)-rhamnosidic linkage, which is difficult to construct with complete stereoselectivity using a direct method. We first examined the glycosylation of 1,2-anhydrorhamnose **114** and 4,6-diol acceptor **115** in the presence of boronic acid catalyst **100** (0.2 equiv.). The reaction proceeded regio- and stereoselectively to provide β(1,4)-rhamnoside **116** in 92% yield as a single isomer. Benzoylation of the 2–OH group in **116**, followed by deprotection of a PMB group, gave disaccharide acceptor **117**. Next, glycosylation of **117** and rhamnosyl donor **118** using TfOH, followed by deprotection of a

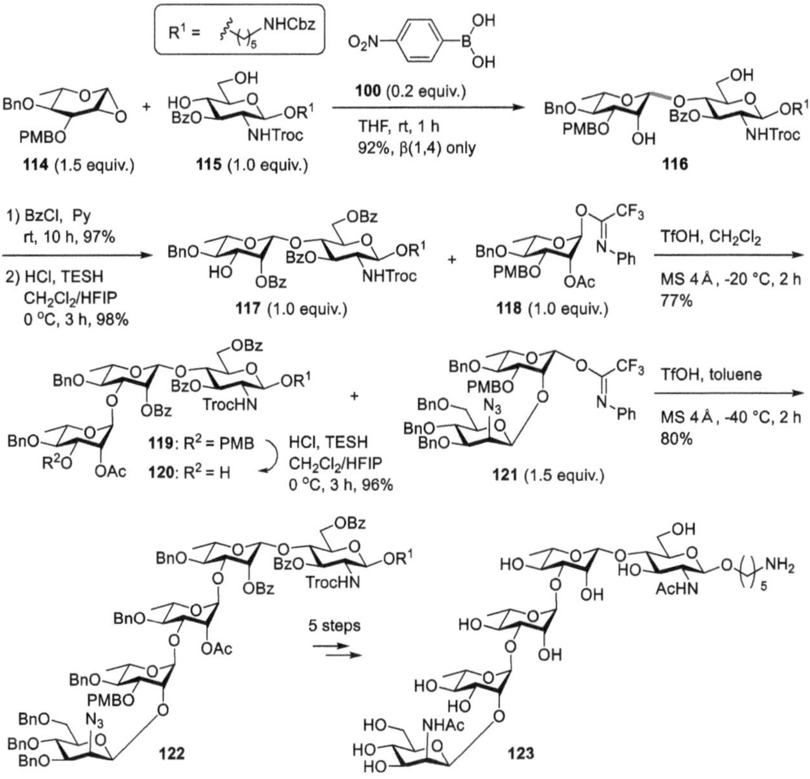

Scheme 8 Synthesis of a pentasaccharide repeating unit of LPS derived from virulent *E. coli* O1.

PMB group in **119**, cleanly afforded trisaccharide acceptor **120** in high yield. Subsequently, TfOH-catalyzed glycosylation of disaccharide donor **121** and **120** proceeded efficiently to provide pentasaccharide **122** in high yield. Finally, conversion of the N_3 group and *N*-Troc group into *N*-Ac groups, followed by removal of the protecting groups, in a total of five steps furnished **123** (Scheme 8).[67]

3.4 Regio- and 1,2-*cis*-stereoselective glycosylations of unprotected glycosides

We applied the BMAD method to unprotected sugar acceptors. Current synthetic schemes are highly dependent on protecting group strategies to control the regio- and stereoselectivity of each glycosylation step, whereas protecting-group-free glycosylation can reduce the number of synthetic steps. We first examined the glycosylation of D-glucal (**124**) with

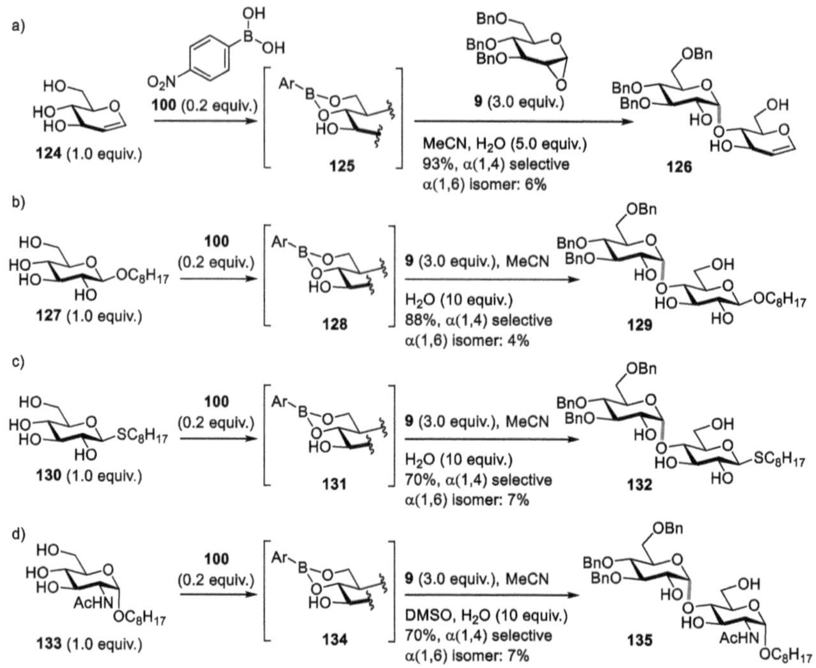

Scheme 9 Regio- and 1,2-*cis*-stereoselective glycosylations of unprotected glycosides.

1,2-anhydroglucose **9** using several boronic acid catalysts under different reaction conditions. After many attempts, we found that the use of boronic acid **100** as a catalyst (0.2 equiv.) supported the efficient glycosylation of **124** and **9** in the presence of an excess of water (5.0 equiv.) in MeCN to provide the corresponding α(1,4)-D-glucoside **126** in 93% yield with high regio- and complete stereoselectivity (Scheme 9A).[61] It should be noted that the use of water as an additive was essential for suppressing the formation of overreacted trisaccharides in this type of BMAD reaction. In addition, we investigated the generality and scope of this reaction using several unprotected sugar acceptors. The results showed that the reaction proceeded smoothly to provide the corresponding α(1,4)-glycosides in good-to-high yields with high regio- and complete 1,2-*cis*-stereoselectivity (Scheme 9B–D).[61]

Next, to better understand the regioselectivity of the present BMAD method, we performed DFT calculations for the formation of α(1,4) and α(1,6) isomers using 1,2-anhydro-3,4,6-tri-*O*-methyl-D-glucose (**136**) and methyl β-D-glucopyranoside (**137**). The results confirmed that both

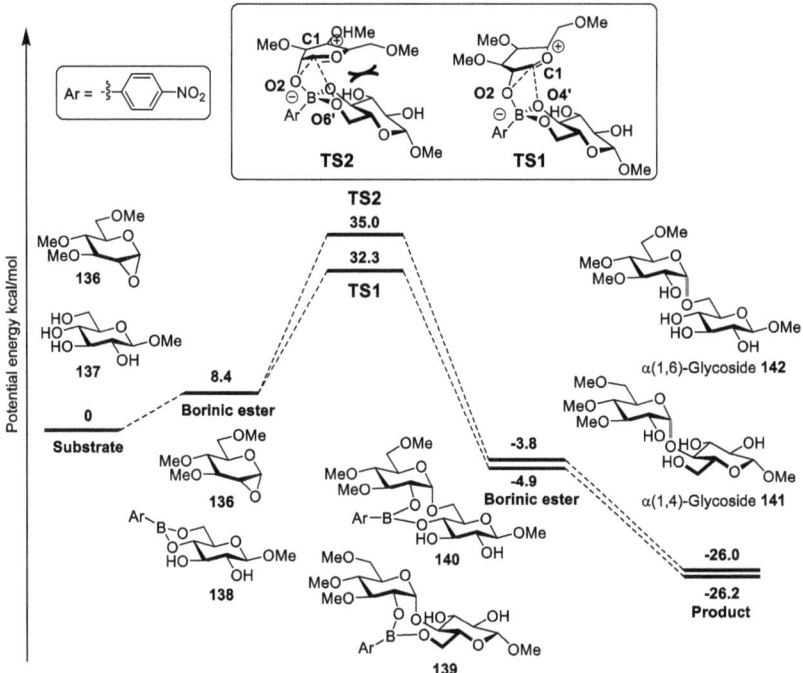

Fig. 7 Potential energy surfaces for BMAD reaction of **136** and **137**.

reactions proceed via an S_Ni mechanism through the transition states TS1 and TS2 to provide $\alpha(1,4)$-glycoside **141** and $\alpha(1,6)$-glycoside **142**, respectively. In addition, the difference in activation energy between TS1 and TS2 was 2.7 kcal/mol, indicating that glycosylation at the 4 position is kinetically favored over the 6 position, in good agreement with the experimental results. Furthermore, the TS structures indicated that regioselectivity is attributed to the overlap between the glycosyl donor and acceptor moieties (Fig. 7).[61]

3.5 Synthesis of branched α-glucan tetrasaccharide unit derived from *Pseudallescheria boydii*

We applied the BMAD method to the synthesis of branched α-glucan tetrasaccharide **150** with minimal protecting groups. Initially, glycosylation of octanol and 1,2-anhydroglucose **9** using borinic acid catalyst **11** gave α-glucoside **143** in 71% yield with complete α-stereoselectivity. Deprotection of the Bn groups provided unprotected sugar acceptor **144**. Glycosylation of tetraol **144** and glycosyl donor **145** using boronic acid catalyst **100** in the presence of water proceeded efficiently to provide desired α(1,4)-glycoside **146** in 72% yield with excellent regio- and stereoselectivity. Subsequently, borinic-acid-catalyzed regio- and stereoselective glycosylation of **9** and tetraol **146** gave α(1,6)-glycoside **147** as a single isomer. Selective deprotection of the PMB groups gave trisaccharide acceptor **148** possessing seven free OH groups. Next, we examined the boronic-acid-catalyzed glycosylation of **9** and heptanol **148** in the presence of water under mild conditions. This reaction also proceeded effectively to give α(1,4)-glycoside **149** in 59% yield (73% BRSM) with excellent regio- and stereoselectivity. Finally, deprotection of the Bn groups furnished branched α-glucan tetrasaccharide **150** (Scheme 10).[61]

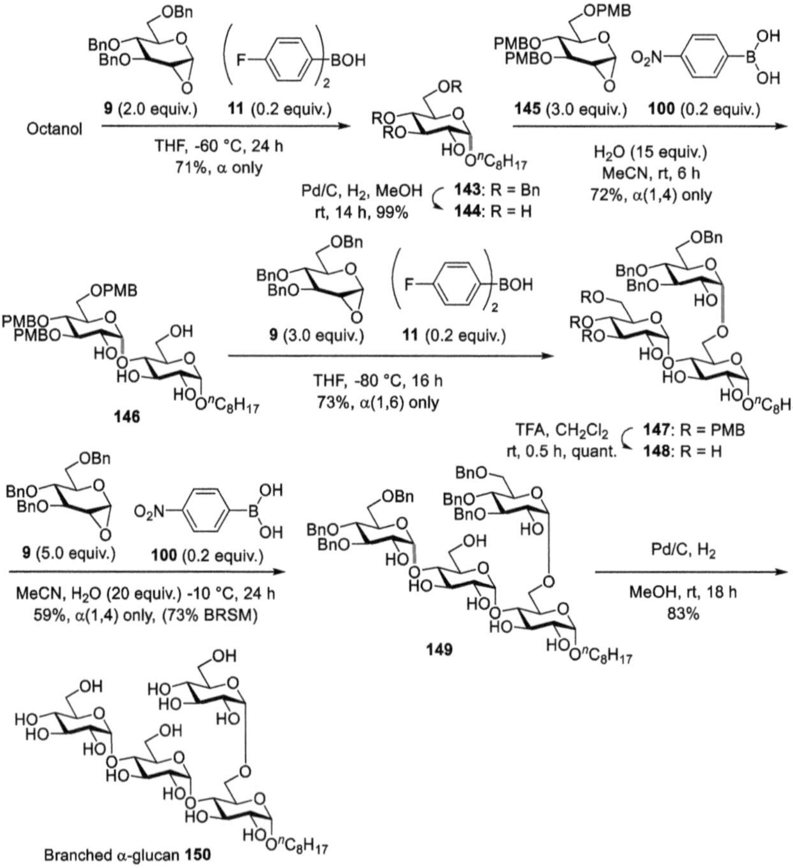

Scheme 10 Efficient synthesis of branched α-glucan tetrasaccharide with minimal protecting groups.

4. Boron-mediated aglycon delivery (BMAD) using a diboron catalyst

4.1 Regio- and 1,2-*cis*-stereoselective glycosylations of *trans*-1,2-diols

Next, we focused on applying the BMAD method to *trans*-1,2-diols on a pyranose ring, and developed the regio- and 1,2-*cis*-stereoselective glycosylation of *trans*-1,2-diols using a diboron catalyst which can bind to *trans*-1,2-diols.[99] We hypothesized that first, diboron **151** reversibly binds to *trans*-1,2-diol **152** to give 1,2-bonded diboron species **153**. Next, diboron **153** activates the 1,2-anhydro glycosyl donor **4** and induces glycosylation to form 1,2-*cis* glycoside **157** with high regio- and stereoselectivity (Fig. 8). Diol exchange reaction between intermediate **156** and *trans*-1,2-diol acceptor **152** regenerates diboron **153** as a catalyst. Thus, this glycosylation reaction is also a regioselective, 1,2-*cis*-stereoselective, catalytic glycosylation method.

To investigate our hypothesis, we first examined the glycosylation of 1,2-anhydroglucose **9** and methyl 4,6-O-benzilidene-α-D-glucopyranoside (**158**) in MeCN at room temperature using catalytic amounts of **151**. The reaction proceeded smoothly to provide α(1,3)-glycoside **159** and α(1,2)-glycoside **160** in 93% total yield with modest regioselectivity (**159/160** = 71:29) and complete α-stereoselectivity (Scheme 11A).[66] Regioselectivity was then investigated using several *trans*-1,2-diol sugar acceptors. Using **161** possessing the cyclic protecting group DTBS as an acceptor, the reaction provided the corresponding α(1,3)-glycoside **162** in high yield with high regioselectivity (**162/163** = 92:8) (Scheme 11B).[66]

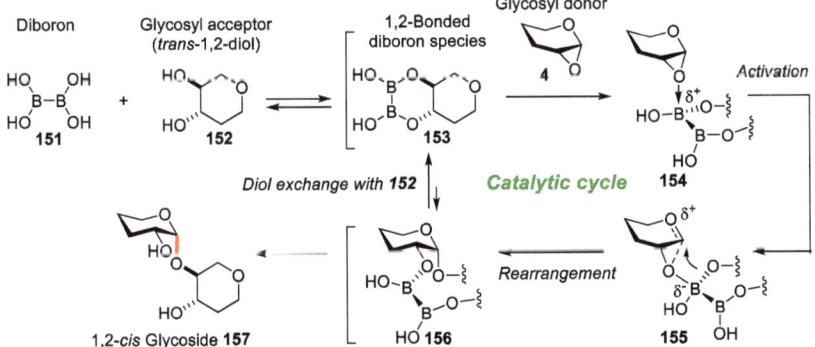

Fig. 8 Concept for BMAD using a diboron catalyst.

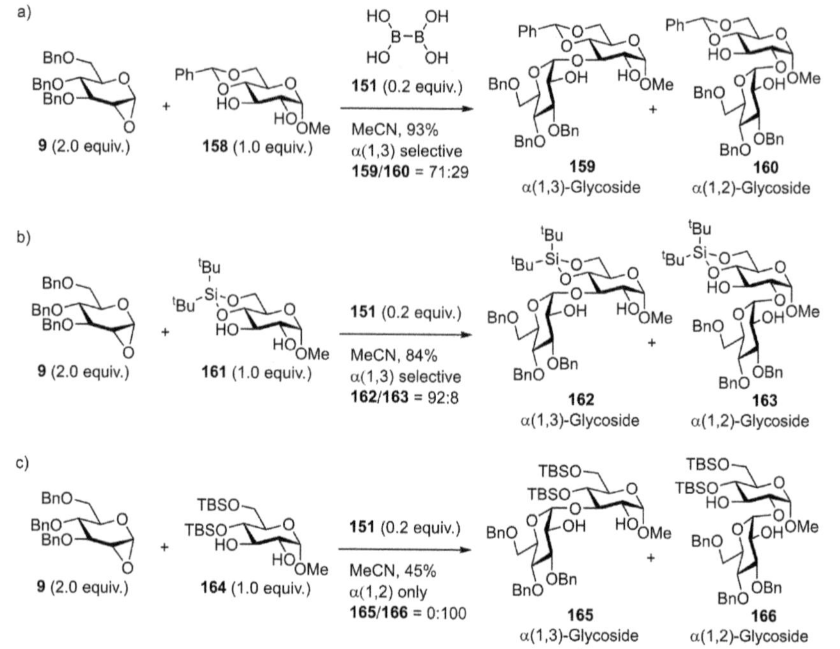

Scheme 11 Regio- and 1,2-*cis*-stereoselective glycosylations of unprotected glycosides.

On the other hand, when **164** possessing acyclic TBS groups was used, α(1,2)-glycoside **166** was obtained as a single isomer in moderate yield (Scheme 11C).[66] These results suggested that steric hindrance of protecting groups at the 4 and 6 positions significantly affects regioselectivity using this BMAD method.

4.2 Synthesis of an α-1,3-glucan pentasaccharide

We applied this BMAD method to the synthesis of an α-1,3-glucan pentasaccharide. Initially, glycosylation of *trans*-1,2-diol acceptor **161** and 1,2-anhydro-3-O-benzyl-4-O,6-O-(di-*tert*-butylsilylene)-α-D-glucose (**167**)[100] using borinic acid catalyst **11** gave α-(1,3)-glucoside **168** in 84% yield (92% BRSM) with complete regio- and stereoselectivity. Deprotection of the Bn groups provided *trans*-1,2-diol acceptor **169**. Two repetitions of the reaction sequence (i) BMAD with **167** and (ii) deprotection of the Bn group provided tetraol **173** in high yield as a single isomer. Subsequently, glycosylation of **173** with **9** proceeded efficiently to afford α-(1,3)-glucoside **174** in 67% yield with good regioselectivity and

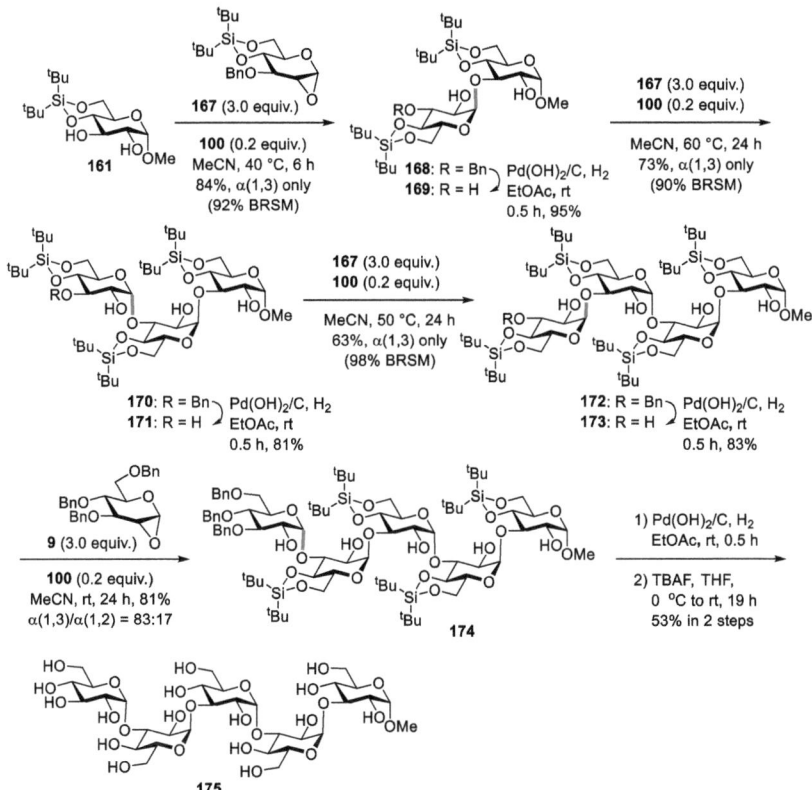

Scheme 12 Efficient synthesis of α-1,3-glucan pentasaccharide.

excellent stereoselectivity. Finally, deprotection of the Bn group and DTBS groups furnished α-1,3-glucan pentasaccharide **175** (Scheme 12).[66]

5. Concluding remarks

This review article provides an overview of our 1,2-*cis*-stereoselective glycosylation methods using organoboron reagents, called boron-mediated aglycon delivery (BMAD). We successfully developed a borinic-acid-catalyzed 1,2-*cis*-stereoselective glycosylation protocol and applied it to the total synthesis of several biologically active glycosides, including GSL-1, acremomannolipin A, and mannosylerythritol lipids (MELs). In addition, we developed a boronic-acid-catalyzed regio- and 1,2-*cis*-stereoselective glycosylation approach and applied it to the direct glycosylation of unprotected sugar acceptors and the synthesis of complex oligosaccharides, including a tetrasaccharide repeating unit of LPS derived from *E. coli* O75, a pentasaccharide repeating unit of LPS derived from virulent *E. coli* O1, and a branched α-glucan tetrasaccharide. Furthermore, we recently developed a diboron-catalyzed regio- and 1,2-*cis*-α-stereoselective glycosylation method for *trans*-1,2 diols and applied it to the efficient synthesis of α-1,3-glucan pentasaccharide. These findings are expected to lead to the development of more efficient glycosylation methods, the total synthesis of biologically active glycosides and highly functional glycosides, and to help further chemical biological research.

References

1. Ernst, B.; Hart, G. W.; Sinaÿ, P., Eds., Vols 1–4; *Carbohydrates in Chemistry and Biology*; Wiley-VCH: Weinheim, 2000.
2. Kamerling, J. P.; Boons, G.-J.; Lee, Y. C.; Suzuki, A.; Taniguchi, N.; Voragen, A. G. J., Eds., Vols 1–4; *Comprehensive Glycoscience: From Chemistry to Systems Biology*; Elsevier: Amsterdam, 2007.
3. Fraser-Reid, B. O.; Tatsuta, K.; Thiem, J.; Coté, G.; Flitsch, S.; Ito, Y.; Yu, B., Eds. *Glycoscience: Chemistry and Chemical Biology*; 2nd ed.; Springer: Berlin, 2008.
4. Biliaderis, C. G.; Izydorczyk, M. S., Eds. *Functional Food Carbohydrates*; CRC Press: Boca Raton, FL, 2007.
5. Chen, J.; Zhu, Y.; Liu, S., Eds. *Functional Carbohydrates: Development, Characterization, and Biomanufacture*; CRC Press: Boca Raton, FL, 2018.
6. Barchi, J. J., Ed., Vols. 1–5; *Comprehensive Glycoscience*; 2nd ed.; Amsterdam: Elsevier, 2021.
7. Toshima, K.; Tatsuta, K. Recent Progress in O-Glycosylation Methods and Its Application to Natural Products Synthesis. *Chem. Rev.* **1993**, *93*, 1503–1531.
8. Jung, K.-H.; Müller, M.; Schmidt, R. R. Intramolecular O-Glycoside Bond Formation. *Chem. Rev.* **2000**, *100*, 4423–4442.
9. Davis, B. G. Recent Developments in Oligosaccharide Synthesis. *J. Chem. Soc., Perkin Trans. 1* **2000**, 2137–2160.
10. Demchenko, D. A. Stereoselective Chemical 1,2-*cis* O-Glycosylation: From 'Sugar Ray' to Modern Techniques of the 21st Century. *Synlett* **2003**, 1225–1240.
11. Toshima, K. Novel Glycosylation Methods and Their Application to Natural Products Synthesis. *Carbohydr. Res.* **2006**, *341*, 1282–1297.
12. Demchenko, A. V., Ed. *Handbook of Chemical Glycosylation: Advances in Stereoselectivity and Therapeutic Relevance*; Wiley-VCH: Weinheim, 2008.
13. Cumpstey, I. Intramolecular Aglycon Delivery. *Carbohydr. Res.* **2008**, *343*, 1553–1573.
14. Zhu, X.; Schmidt, R. R. New Principles for Glycoside-Bond Formation. *Angew. Chem., Int. Ed.* **2009**, *48*, 1900–1934.
15. Mydock, L. K.; Demchenko, A. V. Mechanism of Chemical O-Glycosylation: From Early Studies to Recent Discoveries. *Org. Biomol. Chem.* **2010**, *8*, 497–510.
16. Ishiwata, A.; Lee, Y. J.; Ito, Y. Recent Advances in Stereoselective Glycosylation Through Intramolecular Aglycon Delivery. *Org. Biomol. Chem.* **2010**, *8*, 3596–3608.
17. Kim, K. S.; Suk, D.-H. Effect of Electron-Withdrawing Protecting Groups at Remote Positions of Donors on Glycosylation Stereochemistry. *Top. Curr. Chem.* **2011**, *301*, 109–140.
18. Yang, L.; Qin, Q.; Ye, X.-S. Preactivation: An Alternative Strategy in Stereoselective Glycosylation and Oligosaccharide Synthesis. *Asian J. Org. Chem.* **2013**, *2*, 30–49.

19. Nigudkar, S. S.; Demchenko, A. V. Stereocontrolled 1,2-*cis* Glycosylation as the Driving Force of Progress in Synthetic Carbohydrate Chemistry. *Chem. Sci.* **2015**, *6*, 2687–2704.
20. Williams, R.; Galan, M. C. Recent Advances in Organocatalytic Glycosylations. *Eur. J. Org. Chem.* **2017**, 6247–6264.
21. Jia, X. G.; Demchenko, A. V. Intramolecular Glycosylation. *Beilstein J. Org. Chem.* **2017**, *13*, 2028–2048.
22. Takahashi, D.; Tanaka, M.; Nishi, N.; Toshima, K. Novel 1,2-*cis*-Stereoselective Glycosylations Utilizing Organoboron Reagents and Their Application to Natural Products and Complex Oligosaccharide Synthesis. *Carbohydr. Res.* **2017**, *452*, 64–77.
23. Bennett, C. S., Ed. *Selective Glycosylations: Synthetic Methods and Catalysts*; Wiley-VCH: Weinheim, 2017.
24. Sasaki, K.; Tohda, K. Recent Topics in β-Stereoselective Mannosylation. *Tetrahedron Lett.* **2018**, *59*, 496–503.
25. Panza, M.; Pistorio, S. G.; Stine, K. J.; Demchenko, A. V. Automated Chemical Oligosaccharide Synthesis: Novel Approach to Traditional Challenges. *Chem. Rev.* **2018**, *118*, 8105–8150.
26. Vidal, S., Ed. *Protecting Groups: Strategies and Applications in Carbohydrate Chemistry*; Wiley-VCH: Weinheim, 2019.
27. Báti, G.; He, J.-X.; Pal, K. B.; Liu, X.-W. Stereo- and Regioselective Glycosylation With Protection-Less Sugar Derivatives: An Alluring Strategy to Access Glycans and Natural Products. *Chem. Soc. Rev.* **2019**, *48*, 4006–4018.
28. Ling, J.; Bennett, C. S. Recent Developments in Stereoselective Chemical Glycosylation. *Asian J. Org. Chem.* **2019**, *8*, 802–813.
29. Krasnova, L.; Wong, C.-H. Oligosaccharide Synthesis and Translational Innovation. *J. Am. Chem. Soc.* **2019**, *141*, 3735–3754.
30. Jeanneret, R. A.; Johnson, S. E.; Galan, M. C. Conformationally Constrained Glycosyl Donors as Tools to Control Glycosylation Outcomes. *J. Org. Chem.* **2020**, *85*, 15801–15826.
31. Khanam, A.; Mandal, P. K. Influence of Remote Picolinyl and Picoloyl Stereodirecting Groups for the Stereoselective Glycosylation. *Asian J. Org. Chem.* **2021**, *10*, 296–314.
32. Takahashi, D.; Toshima, K. 1,2-*cis* O-Glycosylation Methods. In *Comprehensive Glycoscience*; Barchi, J. J., Ed.; 2nd ed.; Vol. 2; Elsevier: Amsterdam, 2021; pp. 365–412.
33. Ishiwata, A.; Ito, Y. Synthesis of Docosasaccharide Arabinan Motif of Mycobacterial Cell Wall. *J. Am. Chem. Soc.* **2011**, *133*, 2275–2291.
34. Ishiwata, A.; Sakurai, A.; Nishiyama, Y.; Tsuda, S.; Ito, Y. Synthetic Study and Structural Analysis of the Antifreeze Agent Xylomannan From *Upis ceramboides*. *J. Am. Chem. Soc.* **2011**, *133*, 19524–19535.
35. Shinohara, H.; Matsubayashi, Y. Chemical Synthesis of Arabidopsis CLV3 Glycopeptide Reveals the Impact of Hydroxyproline Arabinosylation on Peptide Conformation and Activity. *Plant Cell Physiol.* **2013**, *54*, 369–374.
36. Kaeothip, S.; Ishiwata, A.; Ito, Y. Stereoselective Synthesis of *Arabidopsis* CLAVATA3 (CLV3) Glycopeptide, Unique Protein Post-translational Modifications of Secreted Peptide Hormone in Plant. *Org. Biomol. Chem.* **2013**, *11*, 5892.
37. Ishiwata, A.; Kaeothip, S.; Takeda, Y.; Ito, Y. Synthesis of the Highly Glycosylated Hydrophilic Motif of Extensins. *Angew. Chem., Int. Ed.* **2014**, *53*, 9812–9816.
38. Corcilius, L.; Hastwell, A. H.; Zhang, M.; Williams, J.; Mackay, J. P.; Gresshoff, P. M.; Ferguson, B. J.; Payne, R. J. Arabinosylation Modulates the Growth-Regulating Activity of the Peptide Hormone CLE40a From Soybean. *Cell Chem. Biol.* **2017**, *24*, 1347–1355.
39. Partridge, K. M.; Bader, S. J.; Buchan, Z. A.; Taylor, C. E.; Montgomery, J. A Streamlined Strategy for Aglycone Assembly and Glycosylation. *Angew. Chem., Int. Ed.* **2013**, *52*, 13647–13650.

40. Walk, J. T.; Buchan, Z. A.; Montgomery, J. Sugar Silanes: Versatile Reagents for Stereocontrolled Glycosylation via Intramolecular Aglycone Delivery. *Chem. Sci.* **2015**, *6*, 3448–3453.
41. Kusumoto, S.; Imoto, M.; Ogiku, T.; Shiba, T. Synthesis of β(1–4)-Linked Disaccharides of *N*-Acetylglucosamine and *N*-Acetylmuramic Acid by Their Direct Condensation. *Bull. Chem. Soc. Jpn.* **1986**, *59*, 1419–1423.
42. Wakao, M.; Fukase, K.; Kusumoto, S. Novel Molecular Clamp Method for Anomeric Stereocontrol of Glycosylation. *Synlett* **1999**, 1911–1914.
43. Wakao, M.; Fukase, K.; Kusumoto, S. Chemical Synthesis of Cyclodextrins by Using Intramolecular Glycosylation. *J. Org. Chem.* **2002**, *67*, 8182–8190.
44. Ziegler, T.; Lau, R. Intramolecular Glycosylation of Prearranged Glycosides. A Novel Tool for Controlling the Reactivity and Anomeric Selectivity of Glycosylations. *Tetrahedron Lett.* **1995**, *36*, 1417–1420.
45. Lau, R.; Schüle, G.; Schwanaberg, U.; Ziegler, T. Intramolecular Glycosylation of Prearranged Saccharides as a Novel Strategy for the Construction of β-L-Rhamnosidic Linkages. *Liebigs Ann.* **1995**, *1995*, 1745–1754.
46. Schüle, G.; Ziegler, T. Synthesis via Prearranged Glycosides of a Tetrasaccharide Fragment Related to the Capsular Polysaccharide of *Streptococcus pneumoniae* Type 27. *Liebigs Ann.* **1996**, *1996*, 1599–1607.
47. Ziegler, T.; Ritter, A.; Hürttlen, J. Intramolecular Glycosylation of Prearranged Glycosides Part 5. α-(1→4)-Selective Glucosylation of Glucosides and Glucosamines. *Tetrahedron Lett.* **1997**, *38*, 3715–3718.
48. Ziegler, T.; Lemanski, G. Synthesis of β-Mannosides via Prearranged Glycosides. *Angew. Chem., Int. Ed.* **1998**, *37*, 3129–3132.
49. Lemanski, G.; Ziegler, T. Prearranged Glycosides, 9. Chemical Synthesis of a Tetrasaccharide Fragment Related to the Exopolysaccharide of *Arthrobacter sp.* CE-17. *Eur. J. Org. Chem.* **2000**, 181–186.
50. Lemanski, G.; Ziegler, T. Synthesis of 4-O-D-Mannopyranosyl-α-D-Glucopyranosides by Intramolecular Glycosylation of 6-O-Tethered Mannosyl Donors. *Tetrahedron* **2000**, *56*, 563–579.
51. Lemanski, G.; Ziegler, T. Prearranged Glycosides, Part 12. Intramolecular Mannosylations of Glucose Derivatives via Prearranged Glycosides. *Helv. Chim. Acta* **2000**, *83*, 2655–2675.
52. Ziegler, T.; Lemanski, G.; Hürttlen, J. Prearranged Glycosides. Part 14: Intramolecular Glycosylation of Non-symmetrically Tethered Glycosides. *Tetrahedron Lett.* **2001**, *42*, 569–572.
53. Lemanski, G.; Ziegler, T. Synthesis of Pentasaccharide Fragments Related to the O-Specific Polysaccharide of *Shigella flexneri* Serotype 1a. *Eur. J. Org. Chem.* **2006**, 2618–2630.
54. Claude, P.; Lehmann, C.; Ziegler, T. Dependency of the Regio- and Stereoselectivity of Intramolecular, Ring-Closing Glycosylations Upon the Ring Size. *Beilstein J. Org. Chem.* **2011**, 7, 1609–1619.
55. Tennant-Eyles, R. J.; Davis, B. G.; Fairbanks, A. J. Peptide Templated Glycosidic Bond Formation: A New Strategy for Oligosaccharide Synthesis. *Chem. Commun. (Cambridge, U. K.)* **1999**, 1037–1038.
56. Tennant-Eyles, R. J.; Davis, B. G.; Fairbanks, A. J. Peptide Templated Glycosylation Reactions. *Tetrahedron: Asymmetry* **2000**, *11*, 231–243.
57. Tennant-Eyles, R. J.; Davis, B. G.; Fairbanks, A. J. Solid Phase Peptide Templated Glycosidic Bond Formation. *Tetrahedron: Asymmetry* **2003**, *14*, 1201–1210.
58. Greenwell, D. R.; Ibnouzaki, A. F.; Warriner, S. L. Peptide-Templated Saccharide Synthesis on a Solid Support. *Angew. Chem., Int. Ed.* **2002**, *41*, 1215–1218.
59. Lin, C.; Jiao, J.; Maisonneuve, S.; Mallétroit, J.; Xie, J. Stereoselective Synthesis and Properties of Glycoazobenzene Macrocycles Through Intramolecular Glycosylation. *Chem. Commun. (Cambridge, U. K.)* **2020**, *56*, 3261–3264.

60. Ikuta, D.; Hirata, Y.; Wakamori, S.; Shimada, H.; Tomabechi, Y.; Kawasaki, Y.; Ikeuchi, K.; Hagimori, T.; Matsumoto, S.; Yamada, H. Conformationally Supple Glucose Monomers Enable Synthesis of the Smallest Cyclodextrins. *Science* **2019**, *364*, 674–677.

61. Tanaka, M.; Nakagawa, A.; Nishi, N.; Iijima, K.; Sawa, R.; Takahashi, D.; Toshima, K. Boronic-Acid-Catalyzed Regioselective and 1,2-*cis*-Stereoselective Glycosylation of Unprotected Sugar Acceptors via S_Ni-Type Mechanism. *J. Am. Chem. Soc.* **2018**, *140*, 3644–3651.

62. Nashida, J.; Nishi, N.; Takahashi, Y.; Igarashi, M.; Hayashi, C.; Takahashi, D.; Toshima, K. Systematic and Stereoselective Total Synthesis of Mannosylerythritol Lipids and Evaluation of Their Antibacterial Activity. *J. Org. Chem.* **2018**, *83*, 7281–7289.

63. Nishi, N.; Sueoka, K.; Iijima, K.; Sawa, R.; Takahashi, D.; Toshima, K. Stereospecific β-L-Rhamnopyranosylation Through an S_Ni-Type Mechanism by Using Organoboron Reagents. *Angew. Chem., Int. Ed.* **2018**, *57*, 13858–13862.

64. Tanaka, M.; Sato, K.; Yoshida, R.; Nishi, N.; Oyamada, R.; Inaba, K.; Takahashi, D.; Toshima, K. Diastereoselective Desymmetric 1,2-*cis*-Glycosylation of *meso*-Diols via Chirality Transfer From a Glycosyl Donor. *Nat. Commun.* **2020**, *11*, 2431.

65. Inaba, K.; Endo, M.; Iibuchi, N.; Takahashi, D.; Toshima, K. Total Synthesis of Terpioside B. *Chem. - Eur. J.* **2020**, *26*, 10222–10225.

66. Tomita, S.; Tanaka, M.; Inoue, M.; Inaba, K.; Takahashi, D.; Toshima, K. Diboron-Catalyzed Regio- and 1,2-*cis*-α-Stereoselective Glycosylation of *trans*-1,2-Diols. *J. Org. Chem.* **2020**, *85*, 16254–16262.

67. Nishi, N.; Seki, K.; Takahashi, D.; Toshima, K. Synthesis of a Pentasaccharide Repeating Unit of Lipopolysaccharide Derived From Virulent *E. coli* O1 and Identification of a Glycotope Candidate of Avian Pathogenetic *E. coli* O1. *Angew. Chem., Int. Ed.* **2021**, *60*, 1789–1796.

68. Kondo, T.; Yasui, C.; Banno, T.; Asakura, K.; Fukuoka, T.; Ushimaru, K.; Koga, M.; Minamikawa, H.; Saika, A.; Morita, T. Self-assembling Properties and Recovery Effects on Damaged Skin Cells of Chemically Synthesized Mannosylerythritol Lipids. *ChemBioChem* **2022**, *23*, e202100631.

69. McClary, C. A.; Taylor, M. S. Applications of Organoboron Compounds in Carbohydrate Chemistry and Glycobiology: Analysis, Separation, Protection, and Activation. *Carbohydr. Res.* **2013**, *381*, 112–122.

70. Dimakos, V.; Taylor, M. S. Site-Selective Functionalization of Hydroxyl Groups in Carbohydrate Derivatives. *Chem. Rev.* **2018**, *23*, 11457–11517.

71. Halcomb, R. L.; Danishefsky, S. J. On the Direct Epoxidation of Glycals: Application of a Reiterative Strategy for the Synthesis of β-Linked Oligosaccharides. *J. Am. Chem. Soc.* **1989**, *111*, 6661–6666.

72. Mori, Y.; Kobayashi, J.; Manabe, K.; Kobayashi, S. Use of Boron Enolates in Water. The First Boron Enolate-Mediated Diastereoselective Aldol Reactions Using Catalytic Boron Sources. *Tetrahedron* **2002**, *58*, 8263–8268.

73. Tanaka, M.; Takahashi, D.; Toshima, K. 1,2-*cis*-α-Stereoselective Glycosylation Utilizing a Glycosyl-Acceptor-Derived Borinic Ester and Its Application to the Total Synthesis of Natural Glycosphingolipids. *Org. Lett.* **2016**, *18*, 5030–5033.

74. Manabe, S.; Marui, Y.; Ito, Y. Total Synthesis of Mannosyl Tryptophan and Its Derivatives. *Chem. - Eur. J.* **2003**, *9*, 1435–1447.

75. Tanaka, M.; Nashida, J.; Takahashi, D.; Toshima, K. Glycosyl-Acceptor-Derived Borinic Ester-Promoted Direct and β-Stereoselective Mannosylation With a 1,2-Anhydromannose Donor. *Org. Lett.* **2016**, *18*, 2288–2291.

76. Chen, Q.; Kong, F.; Cao, L. Synthesis, Conformational Analysis, and the Glycosidic Coupling Reaction of Substituted 2,7-Dioxabicyclo[4.1.0]heptanes: 1,2-Anhydro-3,4-di-*O*-benzyl-β-L- and β-D-rhamnopyranoses. *Carbohydr. Res.* **1993**, *240*, 107–117.

77. Huang, M.; Garrett, G. E.; Birlirakis, N.; Bohé, L.; Pratt, D. A.; Crich, D. Dissecting the Mechanisms of a Class of Chemical Glycosylation Using Primary [13]C Kinetic Isotope Effects. *Nat. Chem.* **2012**, *4*, 663–667.

78. Berti, P. J.; McCann, J. A. B. Toward a Detailed Understanding of Base Excision Repair Enzymes: Transition State and Mechanistic Analyses of *N*-Glycoside Hydrolysis and *N*-Glycoside Transfer. *Chem. Rev.* **2006**, *106*, 506–555.

79. Lee, S. S.; Hong, S. Y.; Errey, J. C.; Izumi, A.; Davies, G. J.; Davis, B. G. Mechanistic Evidence for a Front-Side, S$_N$i-Type Reaction in a Retaining Glycosyltransferase. *Nat. Chem. Biol.* **2011**, *7*, 631–638.

80. Kwan, E. E.; Park, Y.; Besser, H. A.; Anderson, T. L.; Jacobsen, E. N. Sensitive and Accurate [13]C Kinetic Isotope Effect Measurements Enabled by Polarization Transfer. *J. Am. Chem. Soc.* **2017**, *139*, 43–46.

81. Saunders, M.; Laidig, K. E.; Wolfsberg, M. Theoretical Calculation of Equilibrium Isotope Effects Using Ab Initio Force Constants: Application to NMR Isotope Perturbation Studies. *J. Am. Chem. Soc.* **1989**, *111*, 8989–8994.

82. Yamamoto, A.; Yano, I.; Masui, M.; Yabuuchi, E. Isolation of a Novel Sphingoglycolipid Containing Glucuronic Acid and 2-Hydroxy Fatty Acid From *Flavobacterium devorans* ATCC 10829. *J. Biochem.* **1978**, *83*, 1213–1216.

83. Kawahara, K.; Seydel, U.; Matsuura, M.; Danbara, H.; Rietschel, E. T.; Zähringer, U. Chemical Structure of Glycosphingolipids Isolated From *Sphingomonas paucimobilis*. *FEBS Lett.* **1991**, *292*, 107–110.

84. Stallforth, P.; Adibekian, A.; Seeberger, P. H. De Novo Synthesis of a D-Galacturonic Acid Thioglycoside as Key to the Total Synthesis of a Glycosphingolipid From *Sphingomonas yanoikuyae*. *Org. Lett.* **2008**, *10*, 1573–1576.

85. Ndonye, R. M.; Izmirian, D. P.; Dunn, M. F.; Yu, K. O. A.; Porcelli, S. A.; Khurana, A.; Kronenberg, M.; Richardson, S. K.; Howell, A. R. Synthesis and Evaluation of Sphinganine Analogues of KRN7000 and OCH. *J. Org. Chem.* **2005**, *70*, 10260–10270.

86. Sugiura, R.; Kita, A.; Tsutsui, N.; Muraoka, O.; Hagihara, K.; Umeda, N.; Kunoh, T.; Takada, H.; Hirose, D. Acremomannolipin A, the Potential Calcium Signal Modulator With a Characteristic Glycolipid Structure From the Filamentous Fungus *Acremonium strictum*. *Bioorg. Med. Chem. Lett.* **2012**, *22*, 6735–6739.

87. Vidyasagar, A.; Sureshan, K. M. Total Synthesis and Glycosidase Inhibition Studies of (−)-Gabosine J and Its Derivatives. *Eur. J. Org. Chem.* **2014**, 2349–2356.

88. Kitamoto, D.; Akiba, S.; Hiroki, C.; Tabuchi, T. Extracellular Accumulation of Mannosylerythritol Lipids by a Strain of *Candida antarctica*. *Agric. Biol. Chem.* **1990**, *54*, 31–36.

89. Marco, J. A.; Carda, M.; Gonzàlez, F.; Rodriguez, S.; Murga, J. Synthesis of Protected Enantiopure Erythrulose Derivatives. *Liebigs Ann.* **1996**, *1996*, 1801–1810.

90. Nakagawa, A.; Tanaka, M.; Hanamura, S.; Takahashi, D.; Toshima, K. Regioselective and 1,2-*cis*-α-Stereoselective Glycosylation Utilizing Glycosyl-Acceptor-Derived Boronic Ester Catalyst. *Angew. Chem., Int. Ed.* **2015**, *54*, 10935–10939.

91. Nishi, N.; Nashida, J.; Kaji, E.; Takahashi, D.; Toshima, K. Regio- and Stereoselective β-Mannosylation Using a Boronic Acid Catalyst and Its Application to the Synthesis of a Tetrasaccharide Repeating Unit of Lipopolysaccharide Derived From *E. coli* O75. *Chem. Commun. (Cambridge, U. K.)* **2017**, *53*, 3018–3021.

92. Stenutz, R.; Weintraub, A.; Widmalm, G. The Structures of *Escherichia coli* O-Polysaccharide Antigens. *FEMS Microbiol. Rev.* **2006**, *30*, 382–403.

93. Eom, J.-S.; Hwang, B.-Y.; Sohn, J.-W.; Kim, W.-J.; Kim, M.-J.; Park, S.-C.; Cheong, H.-J. Clinical and Molecular Epidemiology of Quinolone-Resistant *Escherichia coli* Isolated From Urinary Tract Infection. *Microb. Drug Resist.* **2002**, *8*, 227–234.

94. Karlowsky, J. A.; Hoban, D. J.; DeCorby, M. R.; Laing, N. M.; Zhanel, G. G. Fluoroquinolone-Resistant Urinary Isolates of *Escherichia coli* From Outpatients Are Frequently Multidrug Resistant: Results From the North American Urinary Tract Infection Collaborative Alliance-Quinolone Resistance Study. *Antimicrob. Agents Chemother.* **2006**, *50*, 2251–2254.

95. Erbing, C.; Svensson, S. Structural Studies on the O-Specific Side-Chains of the Cell-Wall Lipopolysaccharide From *Escherichia coli* O 75. *Carbohydr. Res.* **1975**, *44*, 259–265.

96. Erbing, C.; Kenne, L.; Lindberg, B. Structure of the O-Specific Side-Chains of the *Escherichia coli* O 75 Lipopolysaccharide: A Revision. *Carbohydr. Res.* **1978**, *60*, 400–403.

97. Alberch, L.; Cheng, G.; Seo, S.-K.; Li, X.; Boulineau, P. F.; Wei, A. Stereoelectronic Factors in the Stereoselective Epoxidation of Glycals and 4-Deoxypentenosides. *J. Org. Chem.* **2011**, *76*, 2532–2547.

98. Jann, B.; Shashkov, A. S.; Gupta, D. S.; Panasenko, S. M.; Jann, K. The O1 Antigen of *Escherichia coli*: Structural Characterization of the O1A1-Specific Polysaccharide. *Carbohydr. Polym.* **1992**, *18*, 51–57.

99. Yan, L.; Meng, Y.; Haeffner, F.; Leon, R. M.; Crockett, M. P.; Morken, J. P. On the Carbohydrate/DBU Co-Catalyzed Alkene Diboration: Mechanistic Insight Provides Enhanced Catalytic Efficiency and Substrate Scope. *J. Am. Chem. Soc.* **2018**, *140*, 3663–3773.

100. Wagschal, S.; Guilbaud, J.; Rabet, P.; Farina, V.; Lemaire, S. α-*C*-Glycosides *via syn* Opening of 1,2-Anhydro Sugars With Organozinc Compounds in Toluene/*n*-Dibutyl Ether. *J. Org. Chem.* **2015**, *80*, 9328–9335.

CHAPTER FIVE

Conformationally restricted donors for stereoselective glycosylation☆

Kaname Sasaki* and Nanako Uesaki
Department of Chemistry, Toho University, Funabashi, Japan
*Corresponding author: e-mail address: kaname.sasaki@sci.toho-u.ac.jp

We would like to express our sincere condolences on the loss of Prof. Hidetoshi Yamada. At the conference that KS attended for the first time as a student about 20 years ago, the Yamada group gave a presentation on flipping the conformation of glucoses with bulky silyl groups. Since that time, KS has been fascinated by the chemistry of sugars with unusual conformations. While working on this chapter, Prof. Yamada's foresight and great contribution to synthetic carbohydrate chemistry could be recognized once again. It may not be easy to explore steep paths without this respected chemist, but the authors would like to keep pushing this field forward based on the foundation Prof. Yamada has established.

Contents

☆Dedicated to the memory of Prof. Hidetoshi Yamada.

Advances in Carbohydrate Chemistry and Biochemistry, Volume 82
ISSN 0065-2318
https://doi.org/10.1016/bs.accb.2022.10.005

Abbreviations

3,6-EDB	3,6-[1,1′-(ethane-1,2-diyl)dibenzene-2,2′-bis(methylene)]
aq	aqueous
B	boat conformation
Bn	benzyl
BSM	benzenesulfinyl morpholine
BSP	1-benzenesulfinyl piperidine
BTF	benzotrifluoride (α,α,α-trifluorotoluene)
Bz	benzoyl
C	chair conformation
CD	cyclodextrin (cyclomaltooligosaccharide)
CDA	cyclohexanediacetal
CIP	contact ion pair
DMP	2,6-dimethylphenyl
DMTST	dimethyl(methylthio)sulfonium trifluoromethanesulfonate
DNP	2,4-dinitrophenyl
DTBMP	2,6-di-*tert*-butyl-4-methylpyridine
DTBS	di-*tert*-butylsilylene
E	envelope conformation
GH	glycoside hydrase (glycosidase)
H	half-chair conformation
HAD	hydrogen-bond(ing) (mediated) aglycon delivery
Hex	*n*-hexyl
KIE	kinetic isotope effect
M	molar
MeOTf	methyl triflate (methyl trifluoromethanesulfonate)
NBS	*N*-bromosuccinimide
NIS	*N*-iodosuccinimide
NMR	nuclear magnetic resonance
Pent	pent-4-enyl
Pic	2-picolinyl
Pico	2-picolyl ester
*p***MP**	*p*-methoxyphenyl
RRV	relative reactivity values
S	skew-boat conformation
SSIP	solvent-separated ion pair
TBDPS	*tert*-butyldiphenylsilyl
TBS	*tert*-butyldimethylsilyl
Tf	triflyl (trifluoromethanesulfonyl)

TIPDS	tetraisopropyldisiloxanylidene
TIPS	triisopropylsilyl
Tol	4-methylphenyl
TTBP	2,4,6-tri-*tert*-butylpyrimidine

1. Introduction and the scope of this chapter

1.1 Pyranose ring puckering

Glycoside hydrolases (GH), or glycosidases, are enzymes that accelerate the hydrolysis of glycans. For example, the half-life of cellulose has been extrapolated by Wolfenden et al. to be 4.7 million years at pH 7–14 at ambient temperature,[1] but the hydrolysis reaction proceeds at a practical rate in the presence of the cellulases. A million modules of GHs have been found and classified into more than 170 families based on their amino acid sequence homology and structural features. If you look closely at the mechanism of action, you will notice that GHs utilize a variety of conformations.[2,3]

The conformation of pyranoses is represented by adding the number of atoms that lie out of the plane shared by four or five atoms that constitute the six-membered sugar ring.[4] When the plane is formed by the two pairs of atoms facing each other, the chair (C) is a conformation in which the remaining two atoms are located above or below the plane of the molecule. When the remaining two atoms are located on the same side of the plane, the conformation is called a boat (B) conformation. The plane formed by the four adjacent atoms gives half-chair conformations (H), the plane formed by the three adjacent atoms and one isolated atom gives skew-boat conformations (S), and the plane formed by the five adjacent atoms gives envelope conformations (E). Accordingly, conformations with 32 energetic minima are named (Fig. 1), and GHs have been found to utilize the conformations around $^{4}H_3$, $^{2,5}B$, $B_{2,5}$, or $^{3}H_4$.

It has been found that a 2,4-dinitrophenyl glucoside recognized by the stereoretaining endoglucosidase Cel5A was distorted into a $^{1}S_3$ conformation at the cleavage position, and the covalent intermediate produced was in a $^{4}C_1$ conformation (Scheme 1, eq. 1).[5] The transition state interpolated from the conformations above was estimated to be $^{4}H_3$. This seems to be the most common conformational transition for stereoretaining glycosidases, and the $^{1}S_3$ conformation seems to be the most effective preactivated conformation for β-D-glycosides. As for the stereoretaining endomannosidase Man26A, the Michaelis complex had a $^{1}S_5$ conformation, where the leaving group occupied the pseudoaxial position. and the O-2 substituent occupied

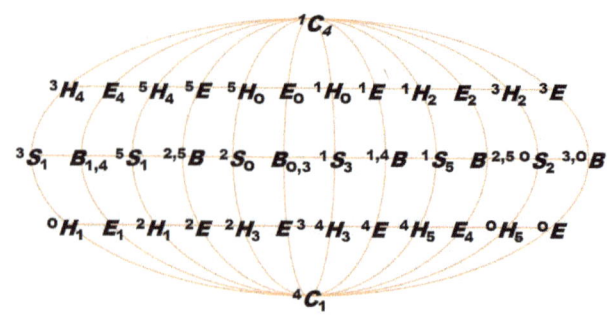

Fig. 1 Spherical mapping of pyranose conformations.

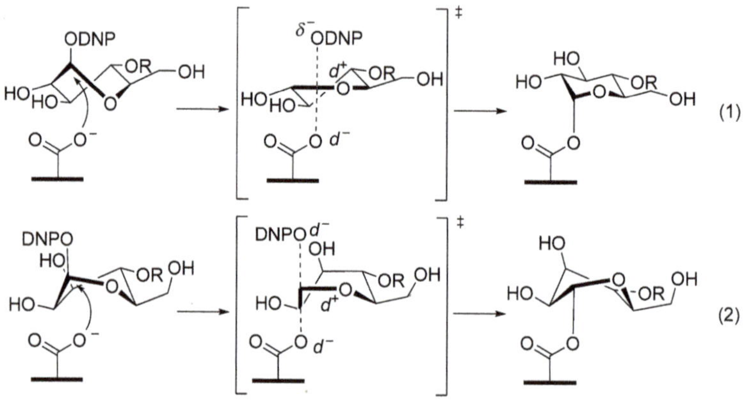

Scheme 1 Substrate distortion along hydrolyzing reaction coordinates by Cel5A (1) and Man26A (2).

the pseudo-equatorial position, thus avoiding the 1,2-syn interaction (eq. 2).[6] Since the covalent intermediate was in an $^{O}S_2$ conformation, the transition state was assumed to be $B_{2,5}$.

In enzymatic reactions, where conformation information on the initial state and the covalent intermediates is available, our understanding of the transition state has been deepened. On the other hand, in chemical glyco-sylation reactions, it had been difficult to know the structure of the interme-diates, although the capture of intermediates by NMR and IR spectroscopy has just begun. However, the knowledge accumulated so far on conforma-tions, reaction rates, and stereoselectivities will greatly help us to infer the transition state. Therefore, the scope of this chapter is to provide examples of chemical glycosylations, especially pyranosylations, in which bulky protecting groups, cyclic protecting groups, or transannular structures were used to fix or change the sterics and thereby control the reaction. Sialylations are discussed by Ando and Komura in Chapter 3 of the preceding volume of this memorial issue and are not included in this chapter.[7]

1.2 Continuum of mechanisms in chemical glycosylation

Glycosylation reactions are generally carried out by nucleophilic substitution reactions using a carbohydrate having a leaving group at C-1 as an electrophile (glycosyl donor) and an alcohol as a nucleophile, although efficient reactions have been reported in reverse order, for example, recently by Li and Zhu.[8] These reactions involve the construction of a new chiral center, and stereoselectivity poses a challenge. The reaction mechanisms are understood to be a continuum of competing mechanisms, including stereo-inversive substitution of a covalent leaving group at the C-1 position of the donor (S_N2), a reaction proceeding from a contact ion pair (CIP) in which the leaving group remains in effect on the stereoselectivity (S_N2-like), and a reaction proceeding from a glycosyl cation intermediate in which the counter anion is separated by a solvent and remains ineffective (S_N1) (Scheme 2). The scheme can be extended to the donor with inverse stereochemistry at C-1 if the leaving group could revert to the glycosyl cation to give an equilibrium. Where in the spectrum the reaction occurs determines the success or failure of the reaction. In particular, in order to control the facial selectivity of nucleophilic addition to glycosyl cations, it would be necessary to control their structure and conformation. In this chapter, we discuss examples in which the reaction mechanisms are selected by controlling the conformation of the substrates.

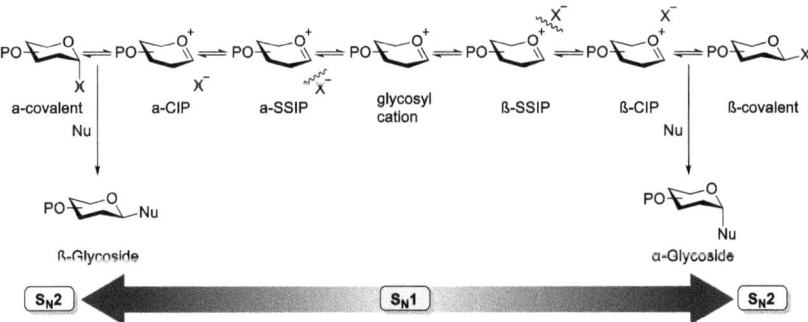

Scheme 2 Continuum mechanisms of glycosylation reactions.

2. Acyclic bulky silyl groups

2.1 Conformational flips induced by bulky silyl groups

Although efforts to twist the pyranose conformations with bulky silyl protective groups had been made before, systematic studies on this subject were attained in 1999 by Yamada and co-workers. They found that the $^1C_4$ conformation of L-rhamnosides with free hydroxy groups at the C-3 and –4 positions became an axial-rich, inverted $^4C_1$ conformation when bulky TBS and TBDPS were introduced into the O-3 and –4 positions, respectively (Scheme 3, eq. 1).[9] Furthermore, glucoses with TBDPS groups at the O-2 and –3 positions also exhibited an inverted $^1C_4$ axial-rich conformation, regardless of the α- or β-anomer (eq. 2).[10] The inversions of xylosides, glucosides, glucosamines, and 2-deoxyglucosides (*arabino*-hexopyranosides) by global introduction of TIPS were also reported by Shuto and co-workers.[11] The conformations inverted by bulky silyl groups have been utilized in the following reactions.

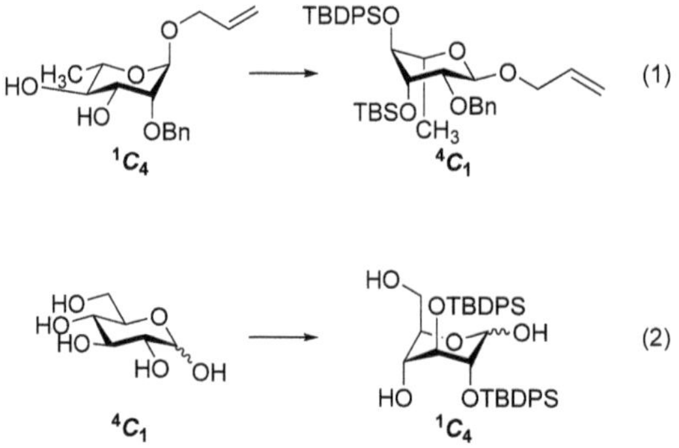

$$ (1) $$

$$ (2) $$

Scheme 3 The conformational inversion of pyranose rings by bulky silyl groups.

2.2 Aryl C-olivosides

In 1996, Suzuki and co-workers reported on the aryl C-glycosylation of olivoses utilizing conformational control to synthesize each anomer (Scheme 4).[12] In general, in aryl C-glycosylation, the adjacent carbons of phenol are replaced by carbohydrates, and the products converge to thermodynamically stable products via quinone methides, mediated by Lewis acids.[13,14] Although the predominant conformation of α-olivosides is the $^1C_4$ conformation, in which the aryl group is oriented equatorially by the steric term, the all-equatorial β-olivoside, which is generated by anomerization via quinone methide, is the most stable. Here, the presence of TBDPS groups at the O-3 and -4 positions stabilized the $^1C_4$ conformation with the two substituents antiperiplanar, thus avoiding the $^4C_1$ conformation with the two bulky substituents in gauche arrangement (Scheme 5). As a result, α-olivosides with an aryl group at C-1 in the equatorial orientation lost the driving force for isomerization to β-olivosides—or β-olivosides gained the driving force for isomerization to α-olivosides—and an unusual α-stereoselectivity arose. Interestingly, they started the synthesis of the olivosides with 6-deoxyglucal having TBDPS groups at the O-3 and -4 positions, and it gave a remarkably flat, half-chair X-ray structure with pseudo-axially oriented O-3 and -4 substituents.

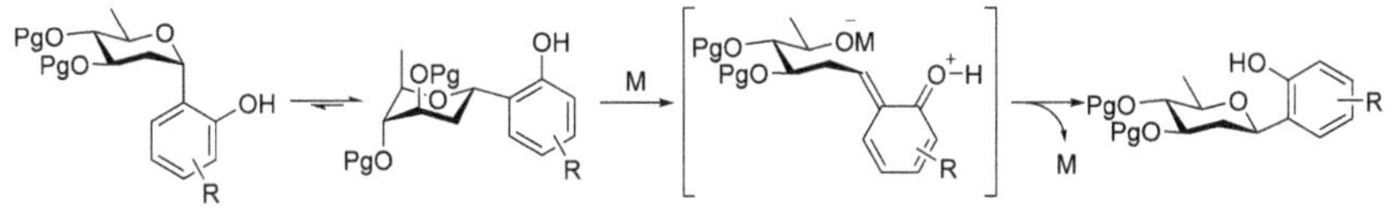

Scheme 4 A general mechanisms based on anomerization of the aryl C-glycosides via quinone methides.

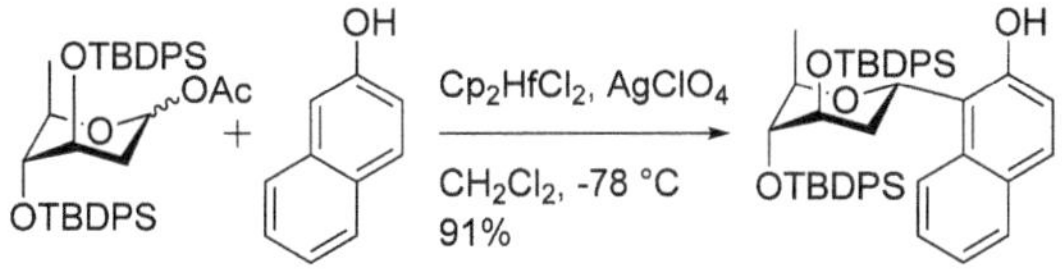

Scheme 5 Aryl α-C-glycosylation enabled by flipped conformations arised from bulky silyl groups.

2.3 S-Glycosides

In 1997, Jackson and co-workers reported that the β-thioglycosides were obtained by ring-opening of the epoxide with three TBS groups. When the reaction was carried out by PhSH in the presence of Et₃N, taking care of the migration of the TBS groups, the S-glycosides had an axial-rich $^1C_4$ conformation both in the crystal structure and in the solution structure in CDCl₃ (Scheme 6).[15] It was speculated that this anomalous conformation was induced by steric hindrance of the two adjacent bulky TBS groups at the O-3 and -4 positions. The silyl group at the O-3 position readily migrated to the O-2 position due to the relaxation of steric hindrance when ring opening was performed under aprotic conditions using PhSLi.

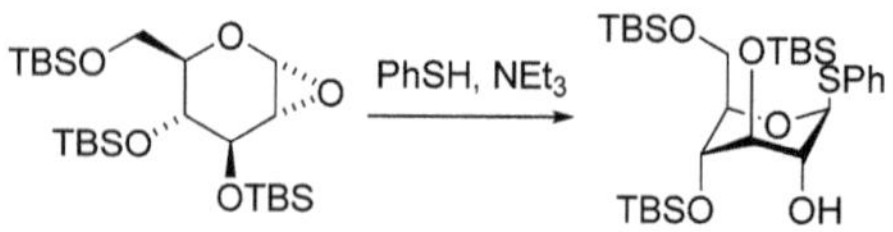

Scheme 6 The anomalous conformation induced by the adjacent TBS groups at the O-3 and -4.

2.4 Allyl C-glycosides

Shuto and co-workers have described the effect of conformations on Sakurai-type C-glycosylation using xylose (Scheme 7).[16] The reaction of a ring-flippable glycosyl fluoride bearing benzyl groups at the O-2, -3, and -4 positions with allyltrimethylsilane proceeded in the presence of $BF_3 \cdot OEt_2$ to give the corresponding C-glycosides with $\alpha:\beta = 2.2:1$ (eq. 1). On the other hand, a glycosyl donor with the O-3 and O-4 positions tethered by Ley's diacetal, which bore a rigid $^4C_1$ conformation, gave $\alpha:\beta > 50:1$ (eq. 2). In addition, a donor equipped with three TIPS groups, which was directed to the $^1C_4$ conformation, reversed the stereoselectivity, and resulted in only the β anomer (eq. 3). The authors explained the stereoselectivity by the interaction of the lone pair n_O of the ring oxygen with the antibonding orbital $\sigma*$ of the bond forming between C-1 and the incoming nucleophile, the so-called pseudo-axial attack; whether it is called a kinetic anomeric effect or not is still controversial.[17]

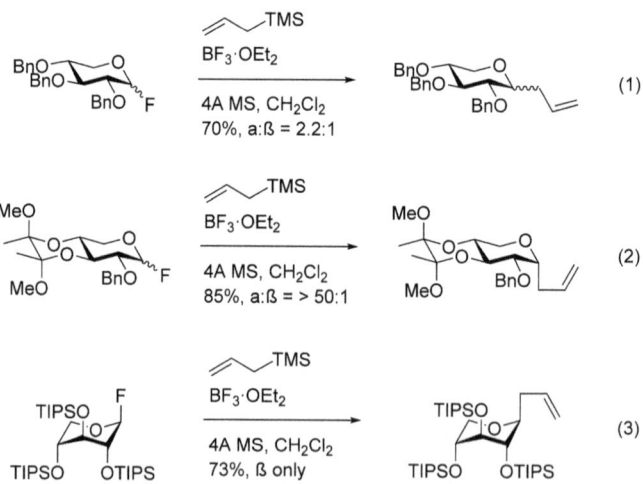

Scheme 7 The conformational effects on allyl C-glycosylation reactions.

2.5 O-Glycosides

C-Glycosylation is considered to follow the S_N1 mechanism, where the reaction proceeds via an exposed glycosyl cation, and the reaction is a useful tool for understanding the stereodirectionality of the intermediate. However, the knowledge from C-glycosylation is not always translated directly to O-glycosylation, because the latter follows a continuum of mechanisms in which the boundary between the S_N1 and S_N2 reactions wanders, depending on the strength of the nucleophile.[18]

Yamada and co-workers performed β-stereoselective glycosylations using ethyl thioglucosides in a $^3S_1$ conformation with three TIPS groups at the O-2, -3 and -4 positions (Scheme 8). When the thioglycosides were activated with MeOTf, the reactions proceeded with β-stereoselectivity predominantly. The protecting group at the O-6 position, the solvent, and the stereochemistry at C-1 of the donor had little effect, and it was concluded that the β-stereoselectivity originated from the skew-boat conformation where the bulky substituent at the O-2 position would stand pseudo-axially.[19] In addition, β-stereoselective glycosylations have been carried out using skew-boat glucosyl donors with another sugar at the O-2 position (Scheme 9).[20] It is interesting to note that the conformational restrictions are utilized here by manipulation of the O-3 and -4 as shown in Scheme 3, which indicates that the bulky substituent at the O-2 position is not necessary.

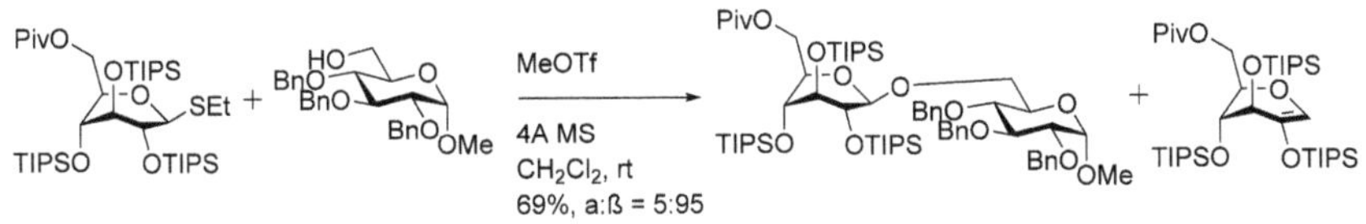

Scheme 8 Pseudoaxial silyl group at O-2 in a skew-boat conformation leads β-stereoselectivity.

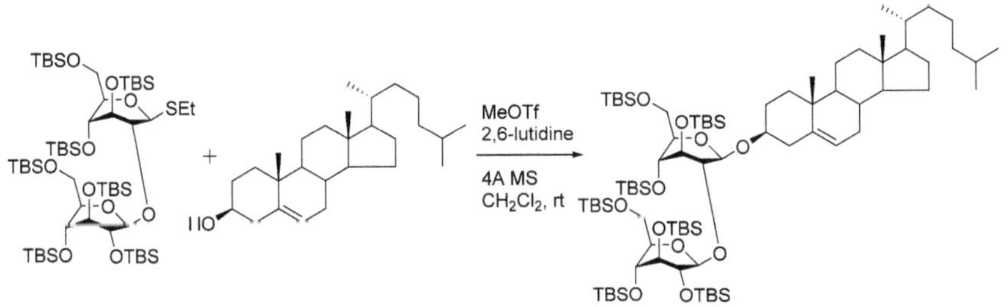

Scheme 9 Skew-boat glucosyl donors with another sugar at the O-2 afford the β-glycosides.

2.6 Effects on reaction rate

2.6.1 Effects of axial substituents

Conformational changes by silyl groups have also been utilized to improve the reactivity of the donors. To understand this, it is helpful to explain that acidic hydrolysis of glycosides was accelerated as the number of oxygen functional groups occupying the axial position increased. This phenomenon has been known for quite a long time, and Bowen and co-workers and Miljkovaić et al. claimed the contribution of the electron term of the axial substituents by MM and ab initio calculations, respectively.[21,22] Then, in 2003, Bols and co-workers experimentally demonstrated that the oxygen function at the 4-position contributed via the electronic term rather than via the steric term (Fig. 2).[23] The relative rate of hydrolysis of the glucoside and the galactoside in 2.0 M HCl aq at 74 °C was 1:5.2. As for those with an extra methyl group at the 4-position, the relative rate was 0.68:4.4, meaning the galactoside was still faster. On the other hand, for a glucoside and a galactoside with a methyl group replacing the hydroxy group at the 4-position, the ratio was 33:21, showing that the glucoside was more reactive than the galactoside. This result rejected the hypothesis of Edward that the relaxation of the steric constriction of the axial substituent should favor the conformational change to produce the glycosyl cation,[24] and revealed the contribution of electronic effects.

In addition, Bols and co-workers synthesized axial-rich glucosides and equatorial-rich galactosides by utilizing the 3,6-anhydro structure, and compared the hydrolysis rates (Fig. 3).[25] The relative hydrolysis rates at 60 °C in 2.0 M HCl aq. were 1.8:446 for the β-glucoside and the axial-rich glucoside with a flipped chair conformation, and 7.2:3.8 for the β-galactoside and the equatorial-rich galactoside with boat conformation. The difference in the

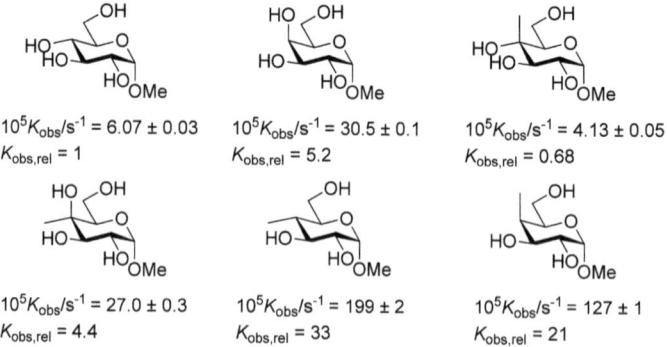

Fig. 2 The oxygen functions at the axial position accelerate the glycoside hydrolysis via electronic term.

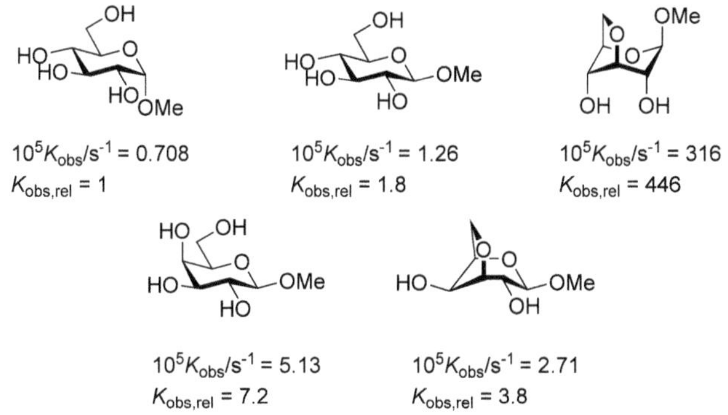

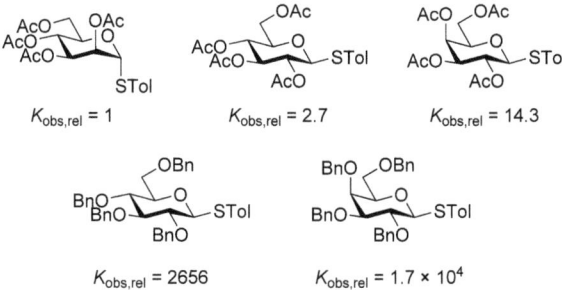

$10^5 K_{obs}/s^{-1} = 0.708$
$K_{obs,rel} = 1$

$10^5 K_{obs}/s^{-1} = 1.26$
$K_{obs,rel} = 1.8$

$10^5 K_{obs}/s^{-1} = 316$
$K_{obs,rel} = 446$

$10^5 K_{obs}/s^{-1} = 5.13$
$K_{obs,rel} = 7.2$

$10^5 K_{obs}/s^{-1} = 2.71$
$K_{obs,rel} = 3.8$

Fig. 3 The axial-rich glucoside is prone to hydrolysis, whereas the equatorial-rich galactoside is tolerant.

$K_{obs,rel} = 1$

$K_{obs,rel} = 2.7$

$K_{obs,rel} = 14.3$

$K_{obs,rel} = 2656$

$K_{obs,rel} = 1.7 \times 10^4$

Fig. 4 The oxygen functional groups at the axial position enhance the reactivity of the glycosyl donors.

rates between the axial-rich glucosides and the equatorial-rich galactosides would be striking and highlight the influence of the oxygen functional group occupying the axial position on the reactivity, even taking the effects of deoxygenation and endocyclic cleavage induced by distortion of the trans-cyclic structure into consideration. The trend is carried over to the reactivity of thioglycoside as glycosyl donors: according to the vast database of relative reactivity values (RRV) determined by Wong et al. (Fig. 4).[26] The RRVs with respect to peracetyl α-thiomannoside are 2.7:14.3 for peracetyl β-thioglucoside and β-thiogalactoside, and 2656:1.7 × 10⁴ for perbenzyl ones, respectively, showing galactosides with an axial substituent are more reactive than all-equatorial glucosides. The trend is also parallel to the pK_a values for piperidinium substituted by hydroxy groups observed by Bols and co-workers (Fig. 5).[27] Piperidinium in the D-gluco configuration, wherein hydroxy group at C-4 occupies the equatorial position, was more acidic than that in the D-galacto-configuration, whose hydroxy group

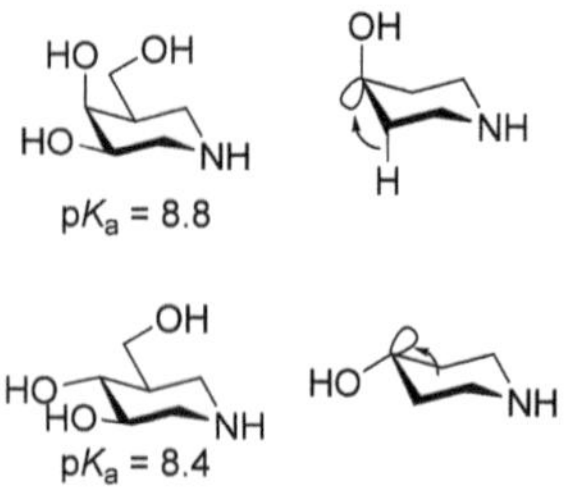

Fig. 5 The equatorial hydroxy group decreases the electron density of the piperidinium.

occupies the axial position. This indicated that the electronegative substituents in the equatorial configuration had a greater effect on reducing the electron density of the ring.

Woerpel and co-workers showed that when there is an electronegative substituent at C-4 of the dioxonium cation, the substituent tended to be pseudo-axially oriented (Scheme 10).[28] The NMR structure in CD_2Cl_2, the X-ray crystal structure, and computational calculations all showed that the pseudo-axial alkoxy group overcame the disadvantage of steric effects and stabilized the C-1 cation, which is parallel to the results of Bols and co-workers in Figs. 3 and 5. Later, NMR analysis revealed a similar trend for the electronegative substituents at C-3.[29]

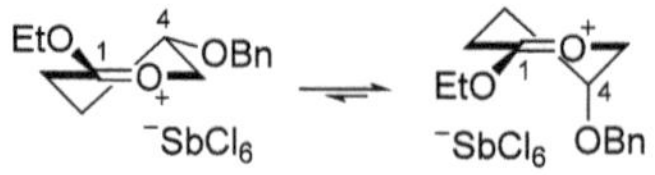

Scheme 10 The conformation with the axial oxygen substituent at C-4 is more stable than that with equatorial one.

2.6.2 Armed effects of bulky silyl groups

In 2007, Bols and co-workers reported the synthesis of disaccharides by super-armed donors with silyl groups, which are more reactive than conventional armed donors with ether protective groups (Scheme 11).[30] Thioglucosides with TBS groups at the O-2, -3, and -4 positions exhibited the $^3S_1$ rather than the $^1C_4$ conformation, regardless of the stereochemistry at C-1, and the oxygen function was pseudoaxial in any case. Glycosylation of the tri-O-benzyl β-thioglucoside proceeded chemoselectively and stereoselectively to give the desired disaccharide. According to Bhat and Gervay-Hague, the silyl group itself has the effect of increasing the reactivity of the donor,[31] but the authors argued that this super-armed effect was due to the conformation. It is expected that these disaccharides can be used as disaccharide donors with armed reactivities.

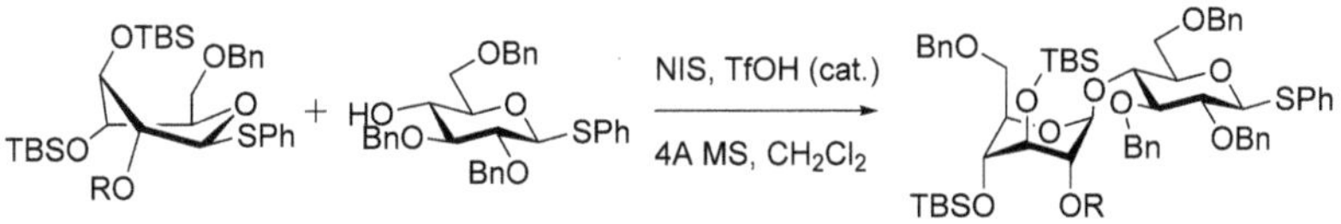

Scheme 11 The super-arming effects of the adjacent bulky silyl groups on the reactivity of the glycosyl donor.

3. Cyclic protective groups

3.1 4,6-Tethered donors

3.1.1 4,6-Tethered mannose

Crich and co-workers have achieved β-stereoselective mannosylation using sulfoxides and thioglycosides bearing a 4,6-O-benzylidene group (Scheme 12).[32–35] It was mechanistically proposed that activation of the sulfoxide with Tf_2O or the thioglycoside with BSP and Tf_2O resulted in the formation of the α-glycosyl triflate due to the anomeric effect, and that stereospecific nucleophilic substitutions with alcohols resulted in the formation of β-glycosides. The 4,6-O-benzylidene group was essential for the stereoselectivity. Variations in the formation of α-triflate include 2-(hydroxycarbonyl)benzyl, benzyl phthalate, and pent-4-enoate explored by Kim and co-workers[36–38] and a hemiacetal by Seeberger and co-workers[39] Trichloroacetimidates,[40] phosphates,[41] N-phenyl trichloroacetimidates,[42] or o-hexynylbenzoates have also been used for β-stereoselective mannosylation with 4,6-O-benzylidene groups.[43] Furthermore, there are variations in the ring-fused protective group at the O-4 and -6 positions, such as boronate,[44,45] cyanobenzylidene,[46] and 1-cyano-2-(2-iodophenyl) ethylidene by Crich and Banerjee,[47] and silylenes by Pedersen, Bols and co-workers[48]

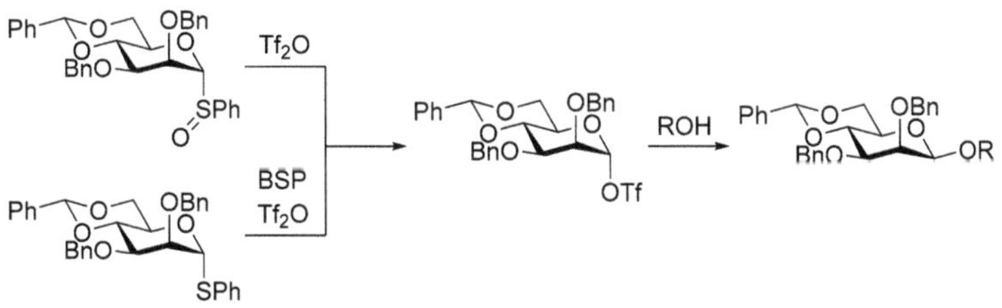

Scheme 12 β-Mannosylation via 4,6-tethered α-triflate.

At least three explanations have been given for the effect of the 4,6-O-benzylidene group. The first is the distortion effect. In 1991, which is earlier than the discovery of the β-mannosylations by Crich and co-workers, Fraser-Reid and co-workers showed that the 4,6-O-benzylidene group had a disarming effect (Scheme 13).[49] The tetra-O-benzyl pent-4-enyl glucosides were oxidatively activated by IDCP in preference to those with 4,6-O-benzylidene groups, resulting in the formation of the desired disaccharide. This has been explained as "torsional disarming," in which the rigid *trans*-decalin skeleton raised the energetic barrier to the glycosyl cation formation with concomitant transition to a half-chair conformation. A similar effect was observed by Ley and Priepke for rhamnosyl thioglycosides with cyclohexanediacetal (CDA) at the O-3 and –4 positions (Scheme 14).[50] 2,3,4-Tri-O-benzyl ethyl thiorhamnoside was activated by IDCP in preference to those inactivated with a fused-ring protective group to give the disaccharide. The latter could also be activated by the more potent NIS-TfOH, which could be used as a disaccharide donor for the second glycosylation.

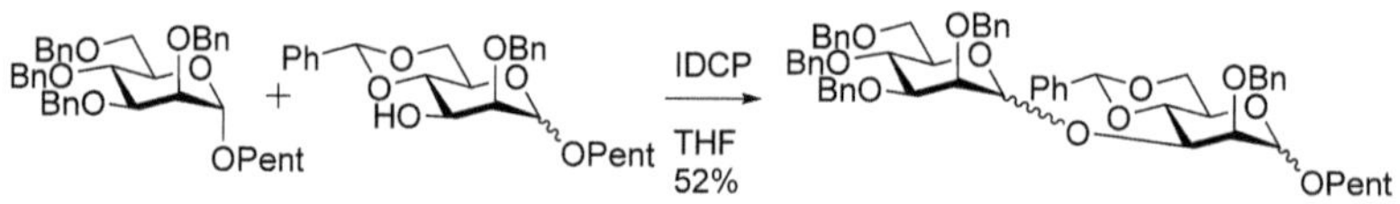

Scheme 13 Disarming effect arisen from 4,6-O-benzylidene group.

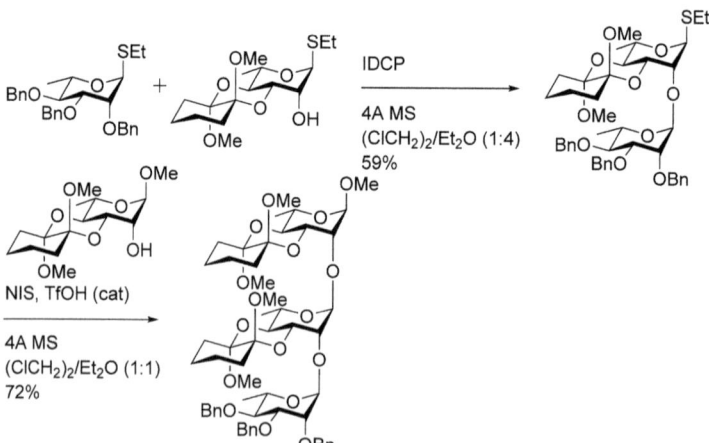

Scheme 14 Disarming effect arisen from 3,4-cyclohexanediacetal group.

The second effect arises from the conformation around C-5–C-6. The oxygen function at the C-6 of mannoses with a 4,6-O-benzylidene group is anchored in the gauche conformation for C-4 and in the trans conformation for O-5, which is called the tg conformation. The conformation lowers the electron density at the O-5 and raises the barrier to glycosyl cation formation. This effect has been clearly shown in the paper by Bols and co-workers in 2004 (Fig. 6).[51] The reaction rates of spontaneous hydrolysis of 2,4-dinitrophenyl (DNP) glucosides with the oxygen functional group at C-6 flexible or immobilized at tg, gt, or gg were examined. The flexible glucoside was found to be the fastest, followed by gg > gt > tg. For galactose, Crich and co-workers have discussed (Fig. 7),[52] and their conclusion was parallel to Bols' model: monocyclic galactosides were the most reactive. For cis-decalin-type galactosides, the reaction rates were gg, gt, and tg conformations in order of increasing reactivity. The reasoning that can be drawn from these two reports is that the effects of the 4,6-O-benzylidene in β-mannosylation is to anchor the O-6 in the tg conformation, which is in antiperiplanar with respect to the ring oxygen, and to maximize the electron-withdrawing effect from the cation center of the glycosyl cation. In this way, the formation of glycosyl cations is retarded. It is clear that the bicyclic structure was indeed responsible for the distortion effect, as the hydrolysis of the monocyclic glycosides were the fastest in both models, but it is also clear that the 4,6-O-benzylidene group raised the energy barrier for the transition to the glycosyl cation.

Yang and Woerpel have reported the computational and NMR structure of a dioxonium cation with an alkoxymethyl group at C-5 (Fig. 8).[29]

Relative hydrolysis rate

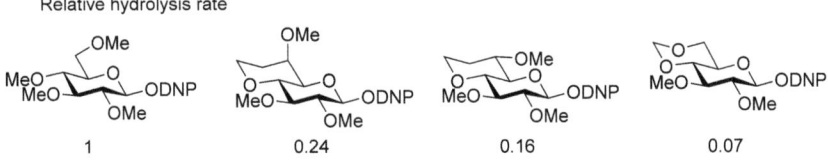

1	0.24	0.16	0.07

Fig. 6 Relative hydrolysis rate affected by the geometry of O-6 in *gluco*-configuration.

Relative hydrolysis rate

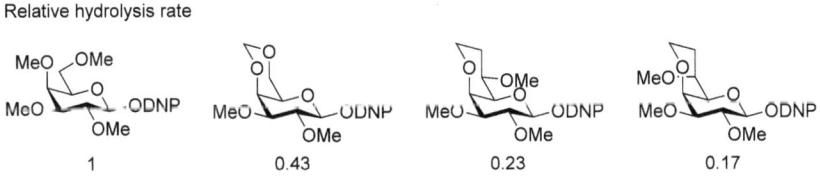

1	0.43	0.23	0.17

Fig. 7 Relative hydrolysis rate affected by the geometry of O-6 in *galacto*-configuration.

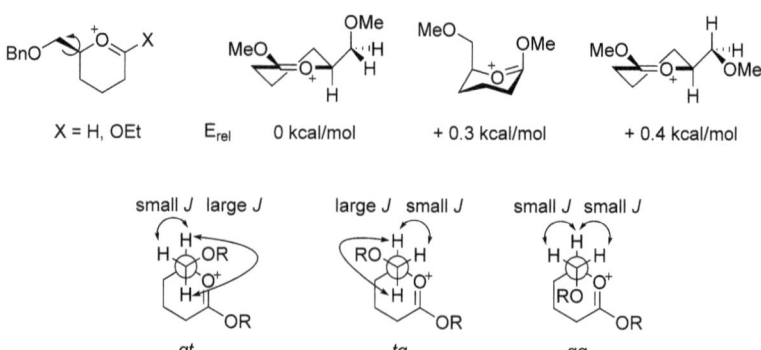

Fig. 8 The tg conformation is the most stable in the dioxonium.

Although the three conformations were close in energy and the solution structures were likely to be mixed, the vicinal couplings possessed by the H-5 were 4.1 and 2.8 Hz (small J-small J), indicating that C-5–C-6–O-6 was in the gg conformation, whether the ring be $^4H_3$ or $^1C_4$. This was consistent with the results of Bols and co-workers: when the electronegative alkoxymethyl group was located near the positive charge of the dioxonium, the conformation should be energetically favored.

The third may be the effect of changing the conformation of the glycosyl cations. The conformation of the glycosyl cations produced from the donors in $^4C_1$ with 4,6-O-acetal seems to be restricted between $^4H_3$ and $B_{2,5}$.[53] The calculations of Hosoya and co-workers showed that the conformation of the contact ion pair (CIP) produced from α-triflate with 4,6-O-formylidene did not converge to a $^4H_3$-like conformation, but exclusively gave a $B_{2,5}$-like conformation,[54] although it was possible to arrive at different scenarios depending on the initial parameters in the calculations.[55] This contrasted with the energetically comparable $B_{2,5}$-like ($^OS_2$/$^3H_2$) and $^4H_3$-like (E_3) conformers obtained from the permethylated α-triflate (Scheme 15).[56,57] In fact, the $B_{2,5}$ conformation has been thought to be β-stereodirectional

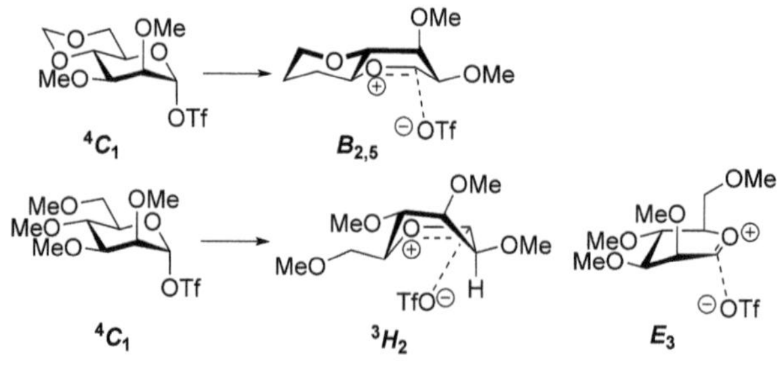

Scheme 15 Plausible conformations of the glycosyl cation with or without 4,6-tether.

in the nucleophilic addition reactions. Experimentally, the primary KIE in the reaction of concerted formation of β-glycosides from α-CIPs in $B_{2,5}$-conformation generated from 4,6-O-benzylidene triflate was in good agreement with the calculations.[58]

3.1.2 4,6-Tethered glucose

Reactions using glucosyl donors with 4,6-O-benzylidene groups might also seem to exhibit stereoinversive β-stereoselectivity, as torsional effects slow down the formation of the glycosyl cation and the stereoelectronic effects of tg conformation destabilize the cation. In fact, however, they afforded exactly the opposite result to mannosyl donors. In their synthesis of the repeating structure of *Streptococcus pneumoniae* capsular polysaccharide, Russo and co-workers have demonstrated a stereoselective α-glycosylation using glucosyl sulfoxides with 4,6-O-benzylidene groups (Scheme 16).[59] Similar reactions have been reported by Crich and Cai.[60] When thioglycosides with 4,6-O-benzylidene were activated, the formation of α-triflate could indeed be confirmed by low-temperature NMR spectroscopy, but when acceptors were added, α-glycosides were the dominant product (Scheme 17).

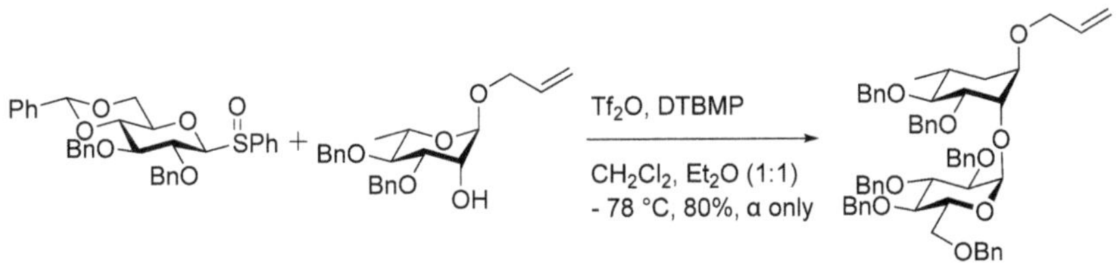

Scheme 16 The glucosyl donors bearing 4,6-O-benzylidene have an α-stereoselective effect.

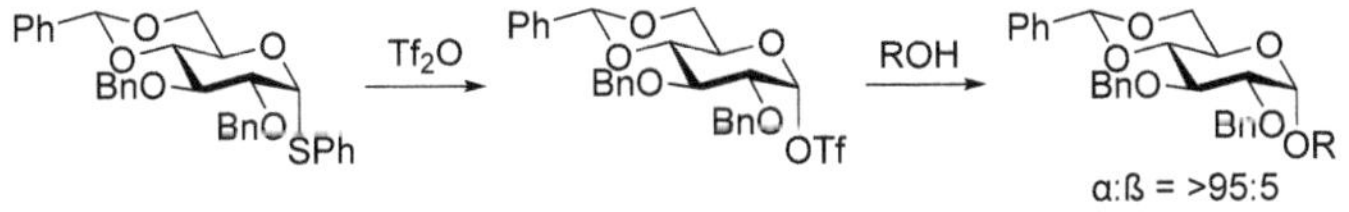

Scheme 17 α-Glucosyl triflate with 4,6-O-benzylidene affords the α-glucoside as the dominant product.

In addition, the silylene tethers were investigated by Heuckendorff and Jensen, showing that the reaction proceeded α-stereoselectively with glucoses, in contrast to β-stereoselectivity with mannoses.[61]

The explanation used to resolve this paradox is the conformational transition to the glycosyl cation. When α-mannosyl triflate, which has an ideal $^4C_1$ conformation, produces glycosyl cations with a $B_{2,5}$ conformation, the

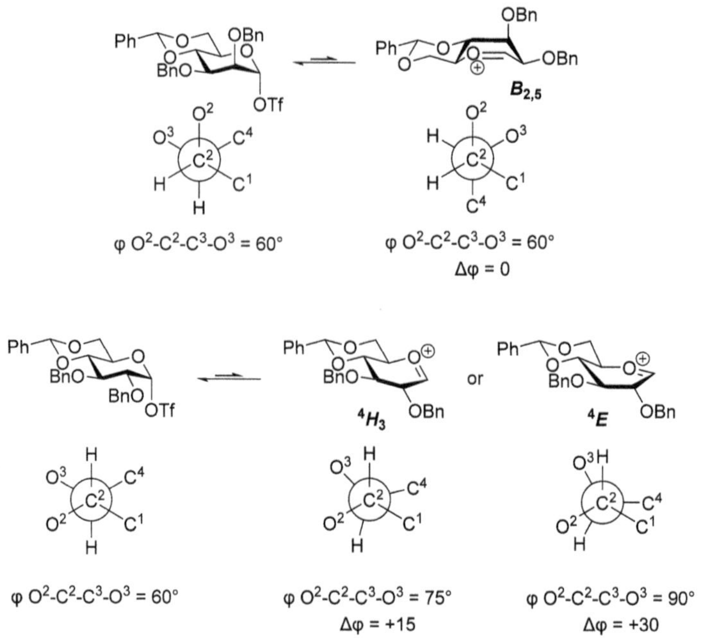

Fig. 9 Steric repulsion between O-2 and -3 in conformational changes.

O-2–C-2–C-3–O3 dihedral angle does not change, whereas when α-glucosyl triflate produces glycosyl cations with a $^4H_3$ or ^4E conformation, the steric repulsion changes in the direction of relaxation (Fig. 9). In other words, the formation of the cation is easier in glucose than in mannose. This was demonstrated by the calculations of Hosoya et al.[54] The contact ion pair (CIP) formed from α-glucosyl triflate with 4,6-O-formylidene had a $B_{2,5}$-like $^1S_5$ conformation, which was similar to that of mannose, but was less stable than SSIP with a $^4H_3$-like^4E conformation. In other words, α-glucosyl triflate seemed to be kinetically and thermodynamically more easily converted to CIP and SSIP than mannosyl one.

Regarding the mechanisms by which the 4,6-O-benzylidene triflate gives the α-glucoside, Crich and co-workers proposed a Curtain–Hammett-type reaction based on the kinetic isotope effect of ^{13}C (Scheme 18).[58] That is, in glucose, where the glycosyl cation was stabilized and the anomeric effect was less effective than in mannose, the covalent α-triflate was in fast equilibrium with the less stable but more reactive β-triflate. Thus, the bimolecular reaction between β-triflate or the resulting β-CIP and the nucleophilic substitution led to the formation of the α-glycosides.

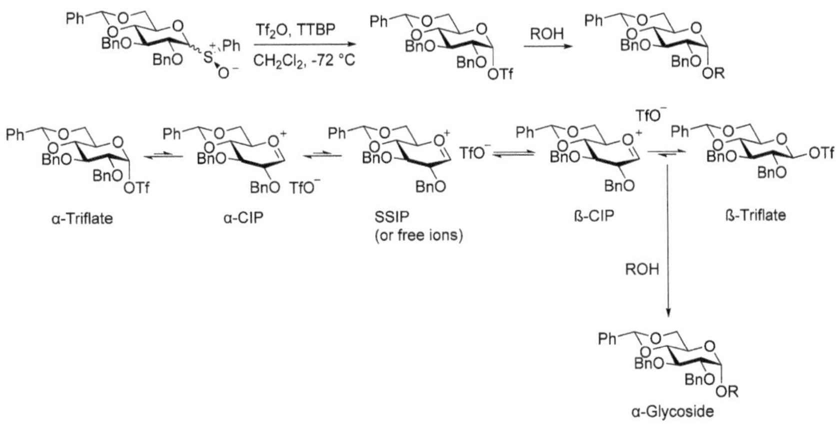

Scheme 18 The less stable but more reactive β-triflate is responsible for the α-stereoselectivity in glucosylation.

Once the effect of the 4,6-O-benzylidene group on α-glucosylation was clarified, it may be natural to expect that it would be extended to the bonds of the 2-amino-2-deoxy-α-D-glucopyranosyl (α-glucosaminyl) moiety, which are more difficult to construct. Actually, this has been attempted by Codée and co-workers,[62] but the results were not as straightforward as expected (Scheme 19). Compared to the 2-O-benzyl group, the 2-azido group is electron-withdrawing and destabilizes the glycosyl cation that is generated from the donor. As a result, the bimolecular mechanisms from a covalent or associative α-triflate to give the β-glycoside could not be ignored, and the stereoselectivity decreased. This tendency was particularly pronounced for the strong nucleophiles. On the other hand, glycosylations of weaker acceptors, whose reactivity was insufficient for S_N2-like reactions, required more electrophilic SSIPs, and the α-stereoselective reactions derived from the $^4H_3/^4E$ glycosyl cation proceeded, as demonstrated by using acceptors with the systematic variation of nucleophilicity.[54,55]

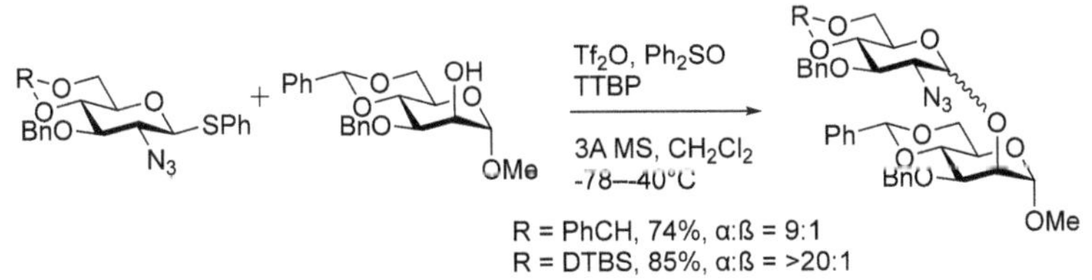

R = PhCH, 74%, α:ß = 9:1
R = DTBS, 85%, α:ß = >20:1

Scheme 19 Silylene tether gives better α-stereoselectivity than benzylidene tether in glucosaminylation.

Codée and co-workers have shown that α-stereoselectivity could be improved by increasing the reactivity of the donors even without changing the acceptor. Thus, replacing the 4,6-O-benzylidene group with a 4,6-O-silylene group increased the reactivity of the donors, and at the same time, slightly improved the α-stereoselectivity.[62]

3.1.3 4,6-Tethered galactose

The mechanisms of the reactions using 4,6-O-benzylidene galactoses have not been discussed as extensively as those of the mannoses and glucoses, but some very complicated but interesting reactions have been reported. In 2001, Wong and co-workers realized an α-stereoselective glycosylation using 4,6-O-benzylidene-containing galactosamine in the synthesis of glycolipids (Scheme 20).[63] The azide group is usually used as a precursor of the nitrogen function at C-2 to avoid neighboring group participation. However, the authors have successfully achieved α-stereoselectivity in the presence of a 2-acetamido group. They speculated that this was because the lone pair from the solvent 1,4-dioxane coordinated to the β-face of the glycosyl cation due to the neighboring group participation of the acetamide function, and the acceptor reacted by kicking it out. On the other hand, Kiso and co-workers reported in 2003 that 4,6-O-benzylidene compounds gave β-glycosides with neighboring group participation of the N-Troc group, whereas replacement of the 4,6-O-benzylidene group with a 4,6-di-O-tert-butylsilylene group resulted in α-stereoselectivity (Scheme 21).[64] This method had a large tolerance for leaving groups,

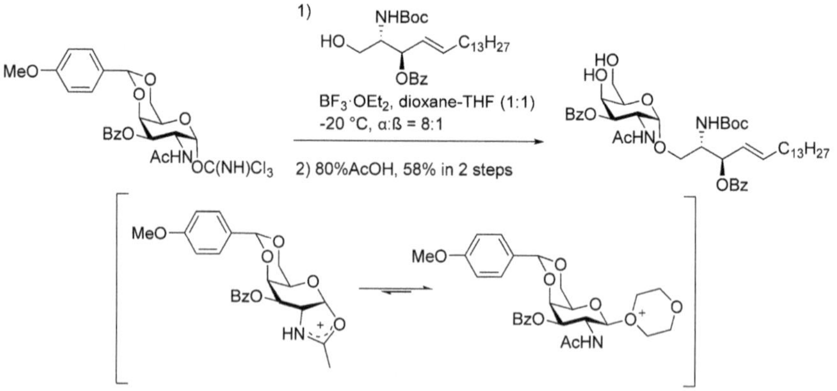

Scheme 20 α-Stereoselective galactosaminylation overcoming the neighboring group participation by 4,6-O-benzylidene.

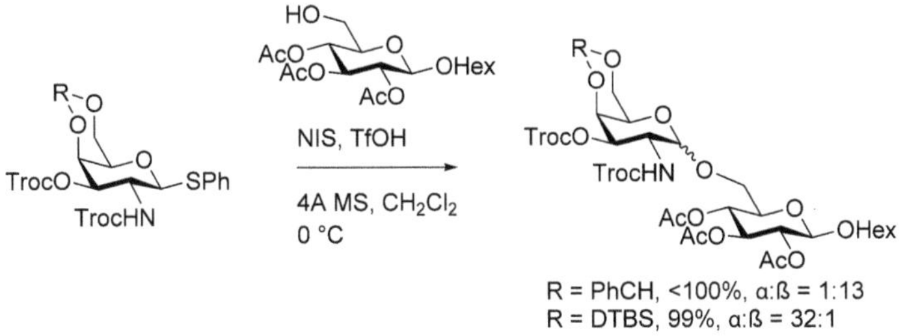

R = PhCH, <100%, α:ß = 1:13
R = DTBS, 99%, α:ß = 32:1

Scheme 21 4,6-Di-O-*tert*-butylsilylene group induces α-stereoselectivity in galactosaminylation.

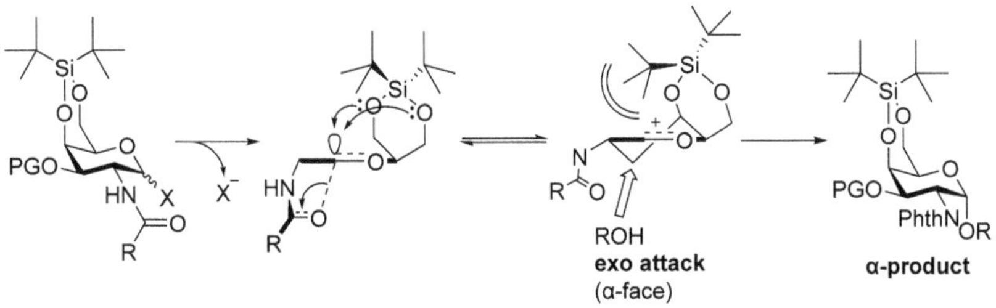

Scheme 22 Postulated mechanism of α-galactosaminylation using DTBS group.

protecting groups at O-2 and –3, and reaction media.[65] Imamura et al. speculated that the stereoselectivity was due to electron donations from O-4 and O-6 in the gg conformation to the oxocarbenium ion, which serve to cleave the neighboring group participation of the N-2 acyl group, and shield the β-face with the bulkiness of the protecting group (Scheme 22).[66]

3.1.4 Hydrogen bond-mediated aglycon delivery with a 4,6-tether

Yasomanee and Demchenko have reported that glycosylation reactions using ethyl thioglucosides with picolinyl or picoloyl groups at the 3-, 4-, or 6-positions proceeded in a syn-stereoselective manner (Scheme 23).[67] That is, when tetra-O-benzyl thioglucosides were activated with DMTST, almost no stereoselectivity was observed (α:β = 1:1.9), whereas replacing the benzyl at O-4 with a picolinyl group resulted in α:β = 5.3:1, and replacing it with a picoloyl group resulted in α:β > 25:1. The reaction was accelerated, and the stereoselectivity was improved, especially when the donor and acceptor were premixed under dilute (5 mM) conditions. Besides, the stereoselectivity was greatly reduced in the presence of DMSO that might have inhibited the hydrogen bonds or strong acids that protonate the pyridine

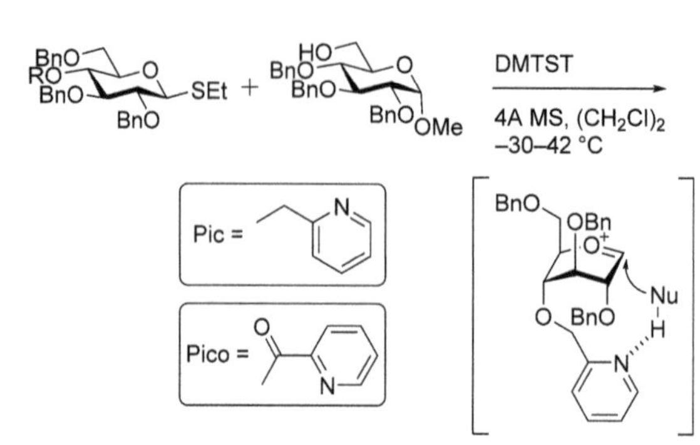

Scheme 23 Hydrogen bond-mediated aglycon delivery.

nitrogen. When the acceptor was silylated to eliminate hydrogen bonds, the stereoselectivity was lost as well. Therefore, a model has been proposed in which hydrogen bonds are formed between the two substrates to define the face that the nucleophile approaches, a phenomenon that is called hydrogen-bonding aglycon delivery (HAD).

Now, what is interesting from the viewpoint of this Chapter is the reaction in which 4,6-O-benzylidene and HAD work together (Scheme 24). Demchemko and co-workers explained that in the case of glucoside synthesis the 4,6-O-benzylidene group reduced the stereoselectivity of HAD, since

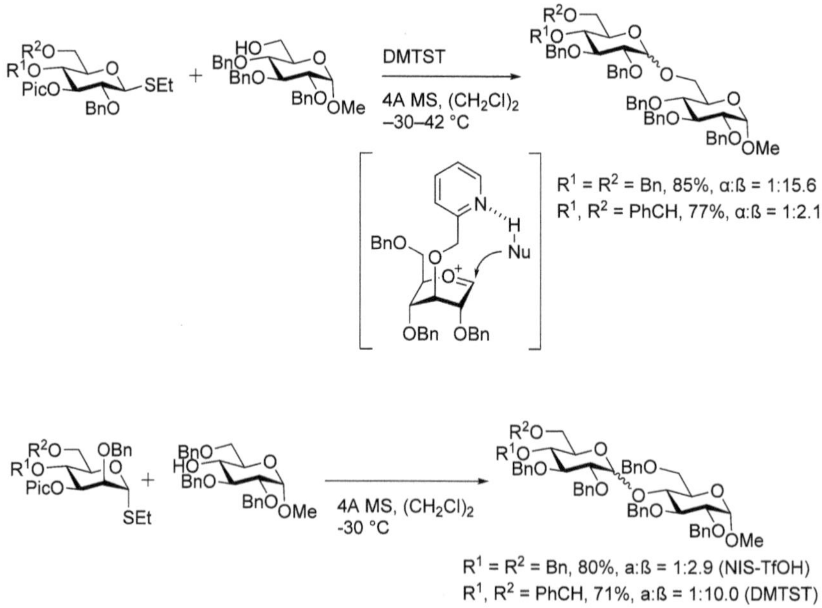

Scheme 24 The effects of 4,6-O-benzylidene group on HAD in *gluco-* and *manno-*series.

it reduced the conformational flexibility needed to generate all-axial glycosyl cations.[67] On the other hand, in the mannosylation of secondary alcohols, the 4,6-O-benzylidene group appeared to improve the stereoselectivity by HAD.[68] This may be related to the fact that a 4,6-O-benzylidene group lowers the barrier to the $B_{2,5}$ conformation, which raises the C-3 substituent to pseudo-axial, as discussed in Section 2.2 of this Chapter. 4,6-O-Benzylidene and HAD work together in the more challenging synthesis of β-mannosamines. Activation of ethyl thioglycosides bearing a picoloyl group at O-3 and an azide at C-2 with NIS-TfOH in dilute solution allowed β-stereoselective mannosaminylation of a wide range of acceptors (Scheme 25).[69]

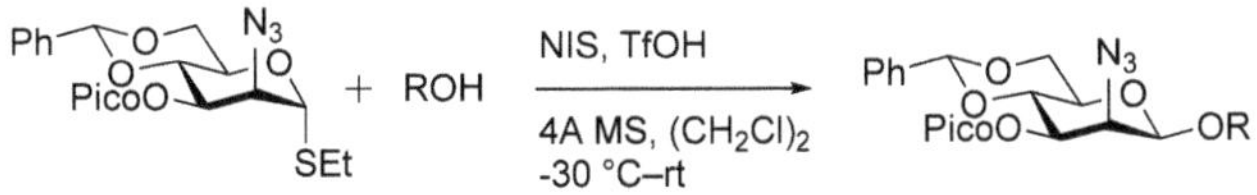

Scheme 25 Collaborative effects of 4,6-O-benzylidene and HAD in β-mannosaminylation.

3.1.5 Remote participation with a 4,6-tether

Crich et al. reported that 4,6-O-benzylidene mannosyl donors with a benzoyl group at O-3, originally intended to stabilize α-triflate to improve β-stereoselectivity by its electron-withdrawing property, exhibited unexpected α-stereoselectivity (Scheme 26, eq. 1).[70] Furthermore, this reaction has been utilized for the synthesis of complex glycans.[71] Demchenko and co-workers in addition, reported that the O-3 benzoyl group exclusively induced α-glycosides even in mannosaminyl donors with an azide at C-2 (eq. 2).[69] Kim and co-workers have reported that the acyl group at O-3 of mannosyl imidate, which did not contain benzylidene and thus had a more flexible conformation, induced α-stereoselectivity, and the stereodirecting effect was larger for acetyl and benzoyl groups and weaker for a p-nitrobenzoyl group, suggesting remote participation as a mechanism other than electron-withdrawing (Scheme 27).[72] An experimental fact that suggested the remote participation was when thiomannoside was activated with BSP-Tf$_2$O in the presence of an O-3 imidoyl group. $^1C_4$ Oxazines were formed by the intramolecular addition of the imidoyl nitrogen to the glycosyl cation. On the other hand, Crich et al. reported that activation of mannosyl sulfoxide bearing a 4,6-O-benzylidene and a 3-O-Boc group did not generate the cyclic carbonate. Glycosides were generated without loss of the Boc group, thus ruling out the possibility of remote participation of the 3-equatorial substituents (Scheme 28).[73]

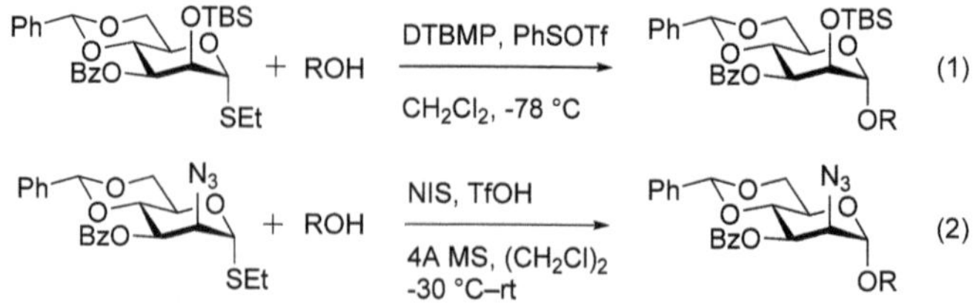

Scheme 26 Benzoyl group at O-3 exhibits α-stereoselectivity in 4,6-O-benzylidene mannosyl donors.

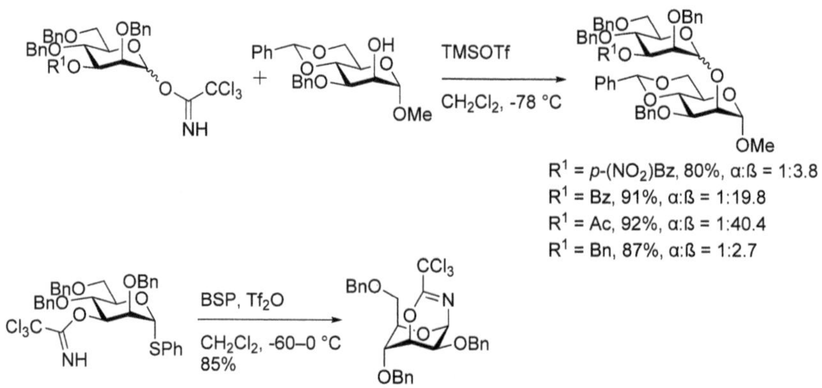

R^1 = p-(NO$_2$)Bz, 80%, α:ß = 1:3.8
R^1 = Bz, 91%, α:ß = 1:19.8
R^1 = Ac, 92%, α:ß = 1:40.4
R^1 = Bn, 87%, α:ß = 1:2.7

Scheme 27 O-3 Imidoyl formed the oxazine when 4,6-O-benzylidene mannosyl donor was activated, suggesting the remote participation enabled α-stereoselective mannosylation.

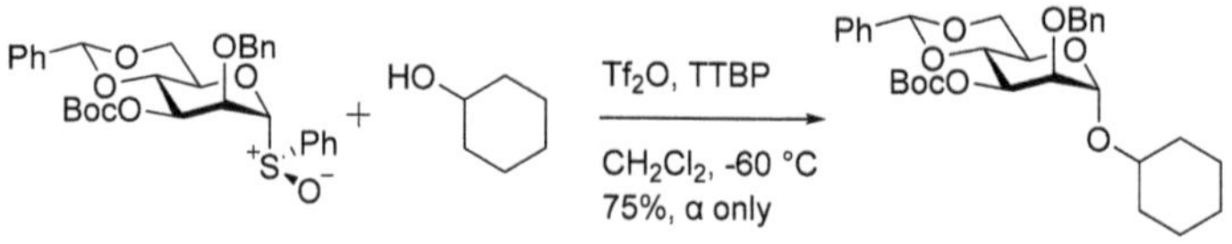

Scheme 28 The mannosyl donor with O-3 Boc group gave the glycoside, ruling out the remote participation.

3.2 3,4- and 2,3-Tethers

3.2.1 The D-manno configuration

Crich et al. have reported that D-mannosyl donors with 3,4–diacetals exhibited a high degree of α-stereoselectivity (Scheme 29, eq. 1).[70] The methoxy group at the axial position was not essential for α-stereoselectivity (eq. 2), and the origin of the α-stereoselectivity was still awaiting a rational explanation. The authors at least ruled out the scenario of destabilization of the α-triflate, which was a dipole parallel to the axial methoxy group. The 3,4-diacetal in mannuronates has also been investigated by Codée et al. (eq. 3), and the formation of α-glycosides took precedence over the β-stereodirecting effect of mannuronates as discussed in Section 4 of this chapter.[74] This indicates that the β-stereodirecting effects of mannuronates come from a glycosyl cation intermediate in a $^3H_4$ configuration, which was not available from the 3,4-tethered sugar donors.

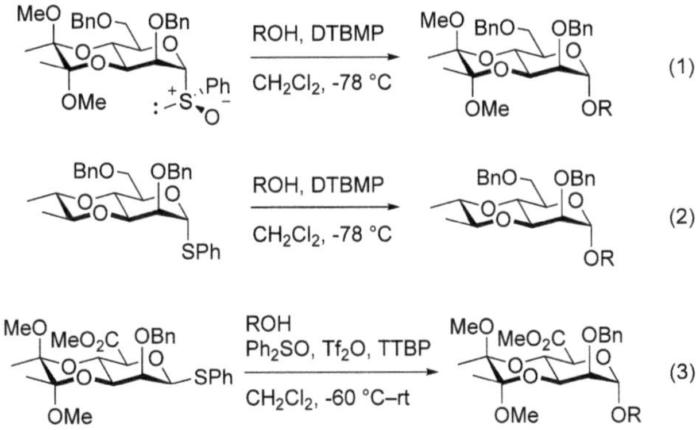

Scheme 29 Six-membered tethers at O-3 and -4 afford α-stereoselectivity in mannosylations.

The glycosylation of 2,3-acetonide-protected mannuronate donors has been reported by Codée, van der Marel, and co-workers and found to be α-stereoselective (Scheme 30).[74] When a β-thioglycoside was activated with Ph_2SO/Tf_2O, the formation of the α-triflate could be observed, but it was of course excluded as a reaction intermediate to explain the α-stereoselectivity.

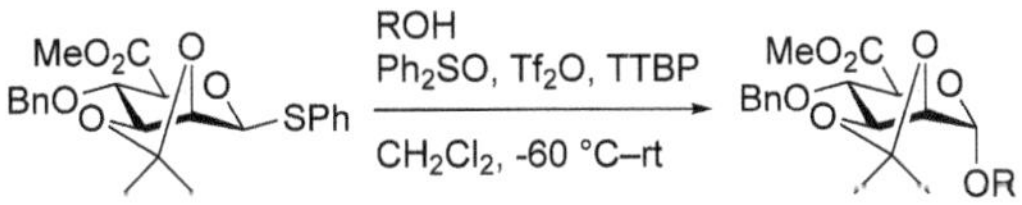

Scheme 30 2,3-O-Acetonide group afford α-stereoselectivity in mannosylation.

As for cyclic 2,3-carbonates, the activation of α-glycosyl bromides with heterogeneous silver salts classically has been used for β-stereoselective rhamnosylation.[75,76] On the other hand, Crich and co-workers reported that this stereoselectivity was specific for insoluble silver salts and became α-stereoselective when activated with soluble silver salts (Scheme 31, eq. 1).[77] The α-stereoselectivities in the homogeneous reactions were also observed when thioglycosides were activated with BSP-Tf$_2$O (eq. 2). They speculated that it was due to the glycosyl cation-like half-chair conformation upon introduction of 2,3-carbonate. On the contrary, the rhamnosyl donors with 3,4-carbonate were β-stereoselective with relatively simple acceptors, because the carbonate was an electron-withdrawing protecting group that did not participate in the reaction (eq. 3). In fact, the stereoselectivity was lost when the carbamate was replaced by acetonide.

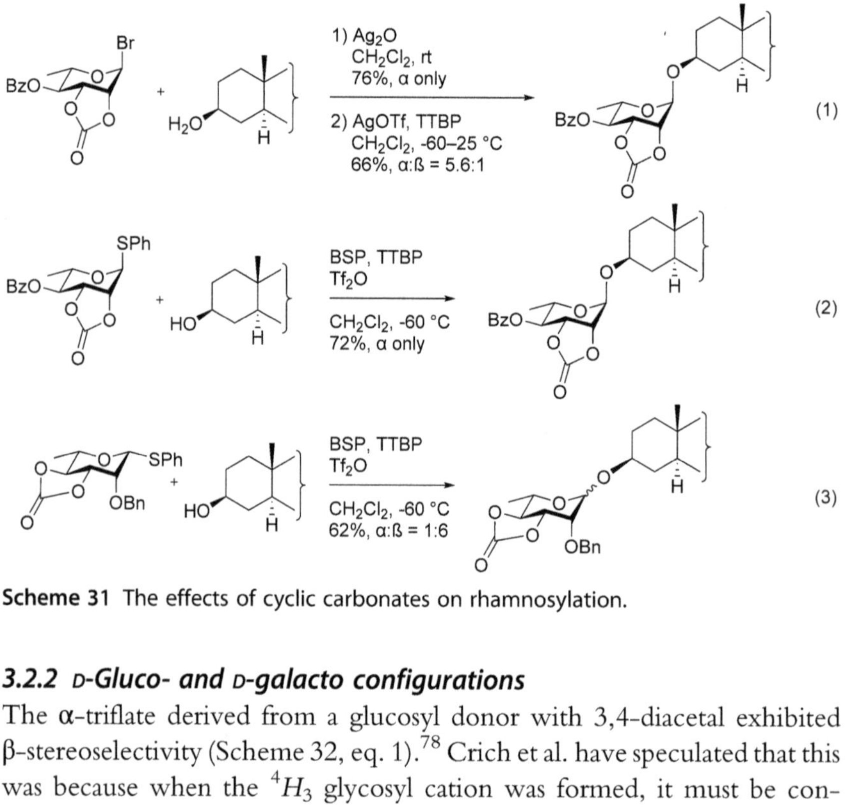

Scheme 31 The effects of cyclic carbonates on rhamnosylation.

3.2.2 D-Gluco- and D-galacto configurations

The α-triflate derived from a glucosyl donor with 3,4-diacetal exhibited β-stereoselectivity (Scheme 32, eq. 1).[78] Crich et al. have speculated that this was because when the $^4H_3$ glycosyl cation was formed, it must be conformationally shifted in a direction that increased the steric repulsion between the C-2 oxygen function and the methoxy group on the α-face. Then, the equilibrium with the glycosyl cation was biased toward the covalent α-triflate, thus contributing to the β-stereoselectivity (eq. 2). In fact, the stereoselectivity was significantly reduced for substrates with the methoxy group removed (eq. 3). When a glycosyl triflate with a 2,3-bisacetal was converted to the glycosyl cation, there was no such interaction, and the stereoselectivity was reduced (eq. 4).

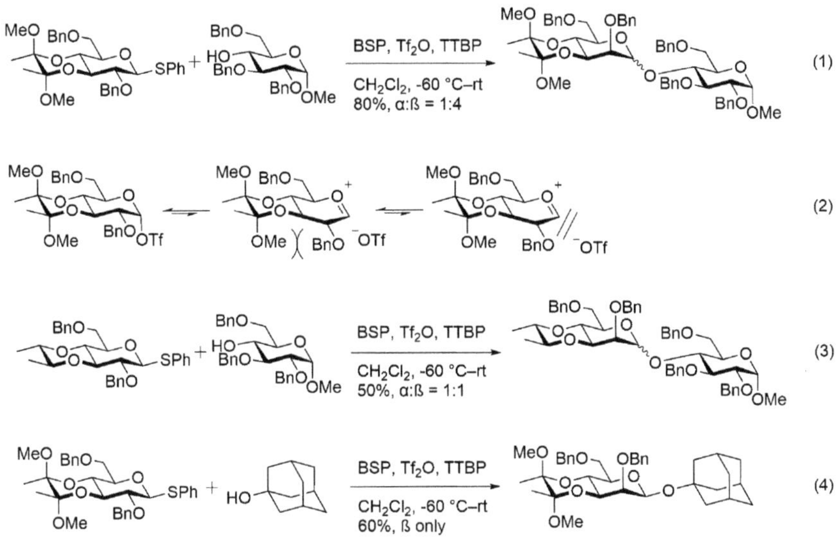

Scheme 32 The effect of the 3,4-diacetal group on the glucosylation.

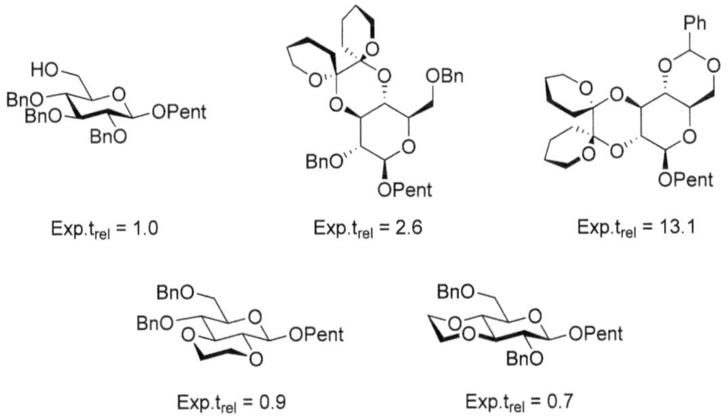

Fig. 10 Relative hydrolysis rate affected by the 3,4- or 2,3-cyclic spirodiacetal protecting group.

In the case of the β-leaving group, Fraser-Reid and co-workers have reported that the 3,4- or 2,3-cyclic spirodiacetal protecting group (dispoke) slowed down the hydrolysis of pent-4-enyl β-glycosides by keeping them out of the flat conformation and destabilizing the intermediate cation (Fig. 10).[79] However, these effects require dispokes, and the ethylene linkages lost their destabilizing effect on the glycosyl cation. In glycosylation reactions, Ishida, Ando and co-workers reported in 2017 that reactions with various 2,3-cyclic protective groups proceeded in a β-stereoselective manner (Scheme 33).[80] This stereoselectivity could be explained by the tethering structure that prevented the pseudo-axial approach of the nucleophile from the α-face when the $^4H_3$ glycosyl cation was generated.

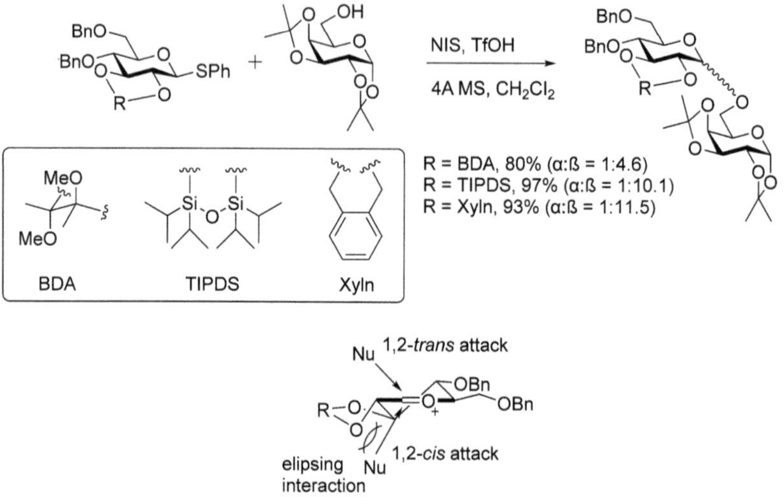

Scheme 33 Stereoselectivities in the reactions using various 2,3-tethers.

The effect of oxazolidinone on α-glucosaminylation has been reported by Kerns and co-workers (Scheme 34, eq. 1).[81] Activation of thioglycosides with 2,3-N,O-oxazolidinone with PhSOTf gave exclusively α-glycosides. Later, Crich and Vinod reported that the C-4 hydroxy group of oxazolidinone-protected glucosamine was an excellent acceptor,[82] and Zhu and Boons reported that this oxazolidinone had a strong disarming effect, and the disaccharide can be synthesized via chemoselective activation against a disarmed donor (eq. 2).[83]

When considering the mechanism of α-stereoselectivity, it is interesting to note that Oscarson and co-workers reported that thioglycosides containing 2-N-acetyl-2,3-N,O-oxazolidinone gave α- or β-glycosides in a highly stereoselective manner, depending on the reaction conditions (eq. 3).[84,85] In the same activation system using NIS-AgOTf, β-glycosides were produced by a short reaction time with a catalytic amount of AgOTf, and α-glycosides were produced by a longer reaction time with an increased amount of AgOTf (up to 0.4 equiv.). Later, Sato, Manabe and co-workers showed that trans-fused pyranosides and oxazolidinones promoted anomerization via endocyclic cleavage in the presence of Lewis

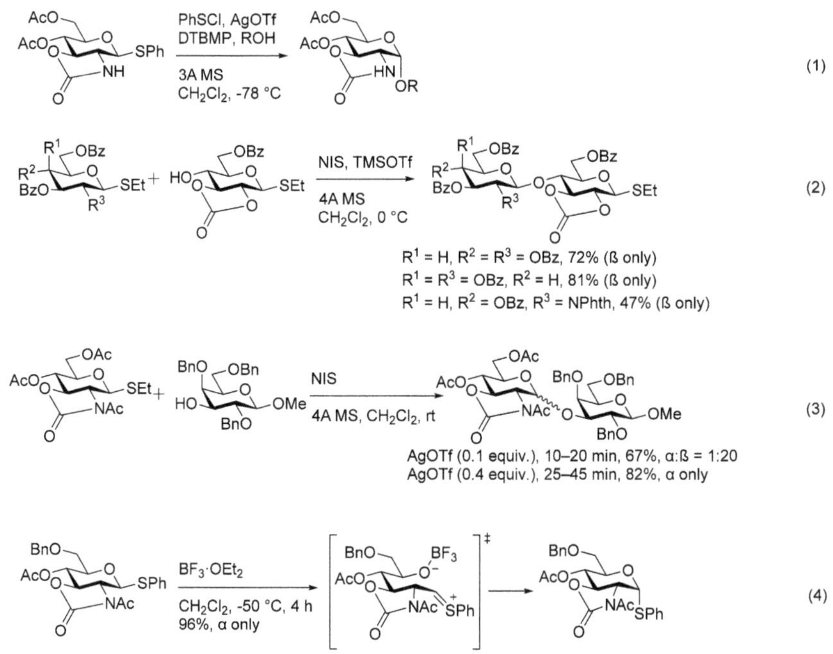

Scheme 34 α-Stereoselective glucosaminylations using the 2,3-oxazolidinone protective group.

acids (eq. 4).[86] Similarly, Ye and co-workers[87] reported that glucosides with 2,3-O-carbonate could also be anomerized in the presence of appropriate Lewis acids to give α-glycosides (Scheme 35).88

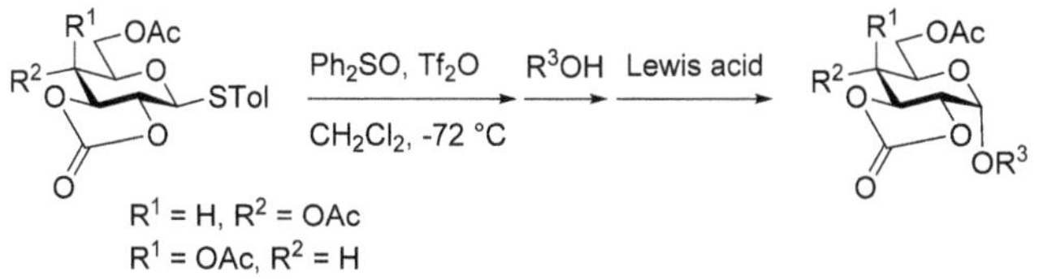

Scheme 35 α-Stereoselective glucosylations via anomerization induced by the 2,3-carbonyl group.

Tatsuta and co-workers reported that when the O-3 and -4 of 2,6-dideoxy thiogalactoside was tethered by acetonide, activation with NBS resulted in high α-stereoselectivity (Scheme 36, eq. 1).[89] The acetonide was essential for this stereoselectivity. Ye and co-workers reported that high α-stereoselectivity was observed when 2-deoxy-1-thioglucoside or -galactoside with O-3 and -4 tethered by carbonate was preactivated with

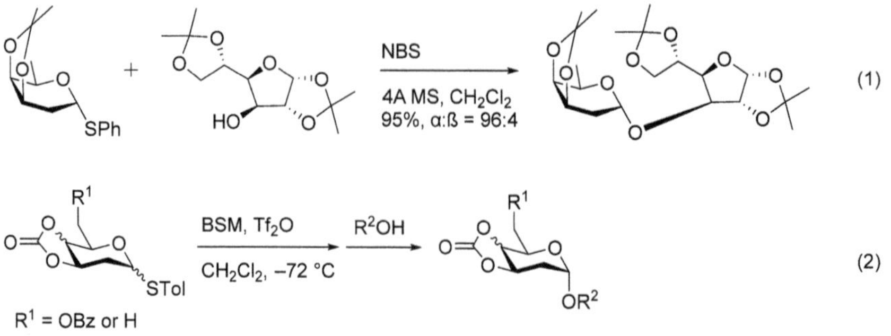
NBS
4A MS, CH_2Cl_2
95%, α:ß = 96:4
(1)
SPh
HO
BSM, Tf_2O
CH_2Cl_2, −72 °C
R^2OH
STol
OR^2
(2)
R^1 = OBz or H

Scheme 36 The effects of the 3,4-tethers in 2,6-dideoxy glycosylations.

benzenesulfinyl morpholine-Tf_2O (eq. 2).[88] This stereoselectivity was not observed when *O*-isopropylidene was used as a tether. It is assumed that the distortion of the conformation contributed to the stereoselectivity in both reactions.

Durham and Roush attempted to synthesize 2-deoxy-β-galactosides assisted by a halide at C-2 and reported that a *cis*-carbonate at O-3 and -4 was essential when using imidates or acetates as donors (Scheme 37).[90] As an intermediate, they postulated a boat-like conformation in which the halogen at C-2 occupied a pseudo-axial position. Indeed, NMR spectral observations of glycosyl cations in superacid by Thibaudeau, Blériot and associates suggested that the glycosyl bromide of 2-bromo-2-deoxy sugars seemed to be in the pseudo-axial position when it contributed to 1,2-trans-β-stereoselectivity.[91]

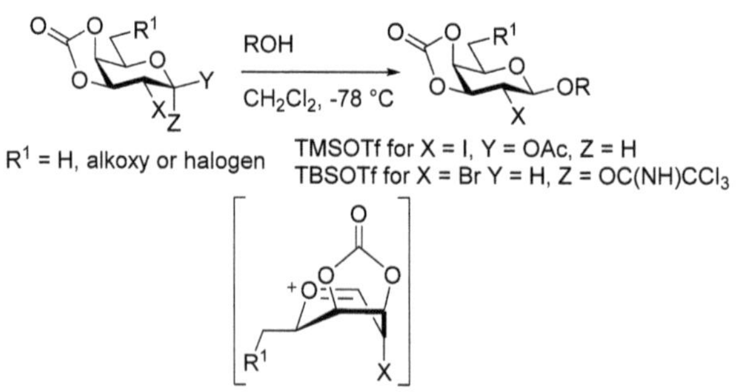
ROH
CH_2Cl_2, -78 °C
OR
R^1 = H, alkoxy or halogen
TMSOTf for X = I, Y = OAc, Z = H
TBSOTf for X = Br Y = H, Z = OC(NH)CCl_3

Scheme 37 3,4-Tethers enhance the participation of the halogen at C-2.

McGarrigle, Galan and co-workers reported that the addition to glucal of 3,4-*O*-disiloxane proceeded with α-stereoselectivity (Scheme 38).[92] This stereoselectivity was not observed when independent silyl groups were

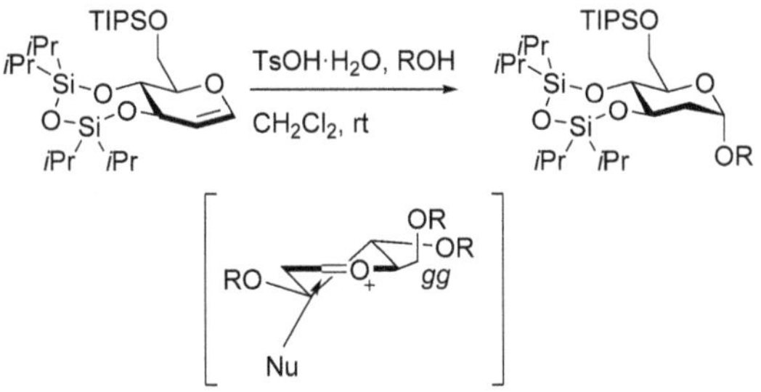

Scheme 38 3,4-Disiloxane induces α-stereoselectivity in addition reaction to the glycal.

introduced at O-3 and -4. The mechanism of this stereoselectivity was attributed to the almost perfect $^4H_3$ conformation of the glycosyl cation based on DFT calculations. The formation of β-glycosides required a glycosyl cation with a $^3H_4$-like or twist-boat conformation, but the former was not possible due to the presence of a cyclic protective group, and the latter was not energetically favorable. Calculations have also shown that the O-6 functional group of the resulting glycosyl cation favored the gg conformation, which was consistent with the arguments of Bols, Woerpel, and Crich.[29,51,52] It was thought that the substituent at C-6 might sterically or stereoelectronically hinder nucleophilic addition to the β-face. In fact, the α-stereoselectivity was somewhat impaired for rhamnal compounds that lacked the oxygen function at C-6.

4. Mannuronates

4.1 Conformations of glycosyl cations generated from mannuronates

In 2006, van der Marel and co-workers reported that the thioglycosides of mannuronates could be activated with Ph_2SO-Tf_2O and then glycosylate acceptors with β-stereoselectivities (Scheme 39).[93] The stereoselectivity tolerated the O-4 acyl group but not the O-3 acyl group (details are given in Section 4.2). The mechanism of this stereoselectivity has been verified computationally and experimentally with simplified substrates, and it has been postulated that the ester at the C-5 position on the oxocarbenium ion was pseudoaxially oriented in the $^3H_4$ conformation with the carbonyl oxygen in proximity to the cation center.[94] Furthermore, the glycosyl triflate of mannuronates with an acyl group at O-4 and an azide group at C-2 by low-temperature NMR studies was shown to be thermodynamically stable in a $^1C_4$ conformation; additionally, the $^1C_4$ conformation of mannolactone

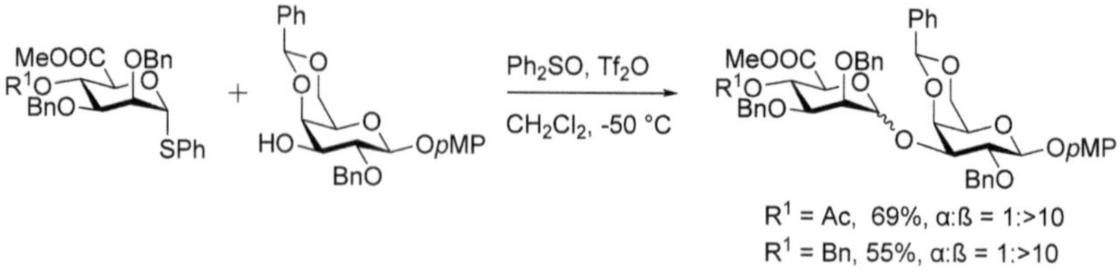

Scheme 39 The thioglycosides of mannuronates afford the β-stereoselectivities when activated with Ph$_2$SO-Tf$_2$O.

with C-1 oxidized, which is structurally analogous to a glycosyl cation, was confirmed by X-ray crystal structural analysis.[95] Nucleophilic pseudo–axial attack at the β-face on the glycosyl cation in $^3H_4$ was considered to be stereoelectronically favorable, leading to the formation of β-glycosides.

Rijs, Boltje and co-workers have reported the gas–phase infrared spectra of the glycosyl cations formed by electrospray ionization of the donors followed by collision-induced dissociation.[96] The tetra-O-methylmannosyl cation showed good agreement with the spectra calculated for the ^3E conformation rather than $^4H_3$. This result was consistent with the stable conformation suggested by Whitfield and co-workers in their calculations (Scheme 40, eq. 1).[97] In the case of the mannuronate cation, on the other hand, the carbonyl oxygen of the ester at C-5 was observed to participate at C-1 (eq. 2). These results suggested that the stereoselectivity of the glycosylation reaction was reversed in mannuronate compared with mannose due to the conformational flip of the intermediates.

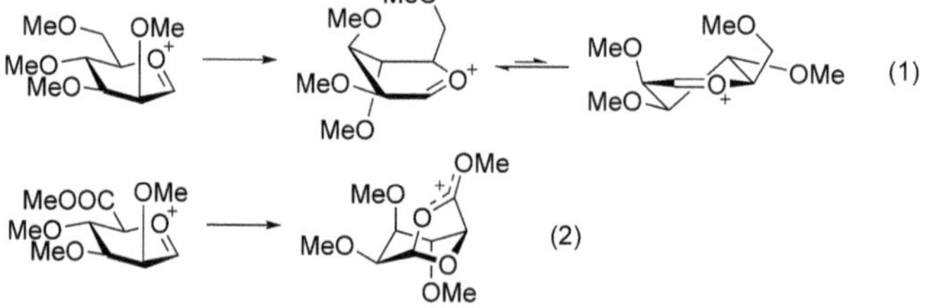

Scheme 40 The gas phase infrared spectra revealed that carbonyl oxygen participated at C-1 of the mannuronate cation.

4.2 HAD and remote participation in mannuronates

In 2021, Demchenko and co-workers demonstrated successfully that the glycosylation reaction using mannuronates as donors could produce both α- and β-glycosides by either HAD or remote group participation (Scheme 41).[98] In contrast to the reaction using a mannosyl donor with a 2-picolyl ester (Pico) group at the O-3, where the stereoselectivity was $\alpha:\beta = 1:3.7$, the reaction using mannuronates showed a strong effect of HAD, increasing the stereoselectivity to $\alpha:\beta = 1:12$. In the reaction using a mannosyl donor with Bz group at O-3, the stereoselectivity was $\alpha:\beta = 1:1.0$,[99] whereas in the reaction using mannuronates, the effect of remote participation was observed, and the stereoselectivity increased to $\alpha:\beta = 25:1$.

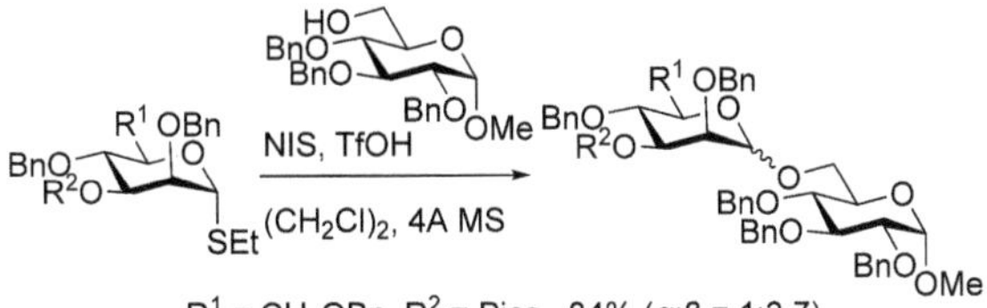

$R^1 = CH_2OBn$, $R^2 = $ Pico, 84% (α:ß = 1:3.7)
$R^1 = COOBn$, $R^2 = $ Pico, 91% (α:ß = 1:12)
$R^1 = CH_2OBn$, $R^2 = $ Bz, 86% (α:ß = 1:1.0)
$R^1 = COOBn$, $R^2 = $ Bz, 92% (α:ß = 25:1)

Scheme 41 HAD and remote participation of O-3 are enhanced in mannuronates when compared with mannoses.

The mechanism of these stereoselectivities can be speculated as follows. The glycosyl cations formed from mannosyl donors are thought to be in the $^4H_3$ conformation, while those from mannuronates are assumed to be thermodynamically favored in the $^3H_4$ conformation, where the cation center is in close proximity to the carbonyl oxygen at the C-6, as discussed in Section 4.1.[94] And in the $^3H_4$ conformation, the substituent at C-3 occupies a pseudo-axial position, which is closer to the reaction center than the pseudo-equatorial position of $^4H_3$. As a result, the effects of HAD and remote participation might strongly emerge (Scheme 42).

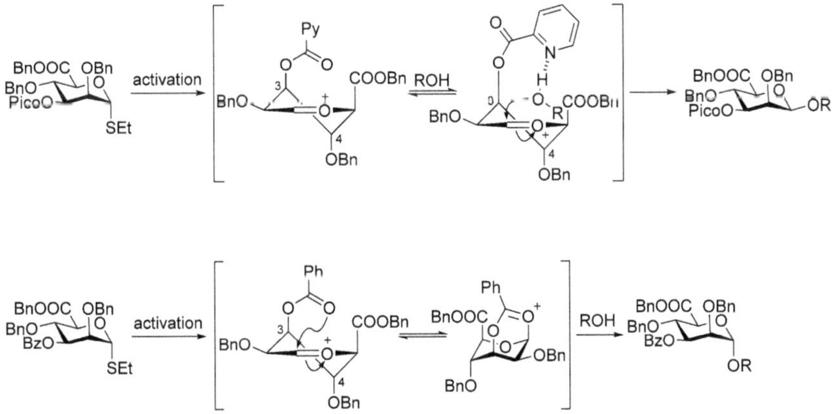

Scheme 42 The effects of the carbonyl oxygen at C-6 on HAD and remote participation.

A seminal discussion of remote participation in mannuronates has been presented by Rijs, Boltje and co-workers using infrared spectra data of carbocations supplemented with quadrupole ion traps (Scheme 43).[96] For mannuronate cations with an acetyl group at O-4, the spectra were a mixture of both the participation of the ester oxygen at the C-5 and the participation of the acyl group at O-4. Both interactions contributed to the enhancement of β-stereoselectivity, the former by directing the glycosyl cation to the β-stereodirecting $^3H_4$ conformation and the latter by shielding the α-face of the reaction center.

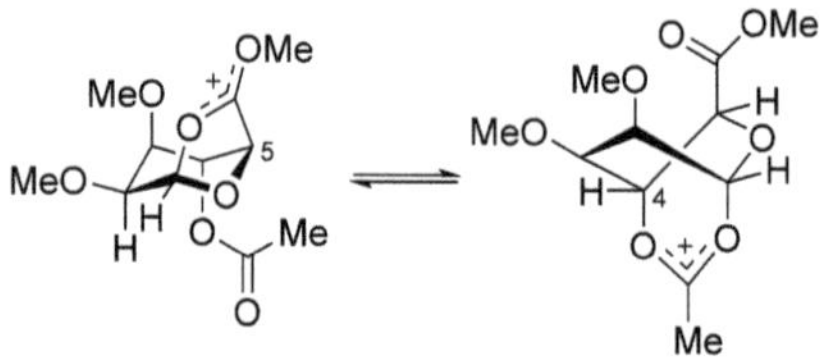

Scheme 43 The gas phase infrared spectra revealed that either carbonyl oxygen at C-6 or acyl group at O-4 could participate at C-1 of the mannuronate cation.

5. Transannular structures

5.1 3,6-Tethered donors

5.1.1 3,6-Lactones

van der Marel and co-workers reported an α-stereoselective reaction using galacturono-3,6-lactone (Scheme 44).[100] NMR measurements revealed that the conformation was inverted to $^1C_4$ conformation,[101] but in some cases the α-stereoselectivity was significantly reduced, the mechanism of which has not been discussed.

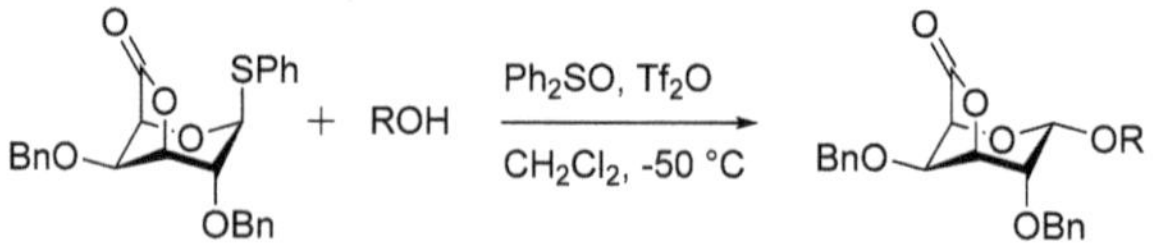

Scheme 44 α-Stereoselective reaction using galacturono-3,6-lactone.

Boltje and co-workers reported an interesting reaction using donors with a mannurono-3,6-lactone structure.[102] To understand the background to this reaction, we should start the story with the 1,2-cis-α-glycosylation utilizing an asymmetric auxiliary at O-2 reported by Boons and co-workers

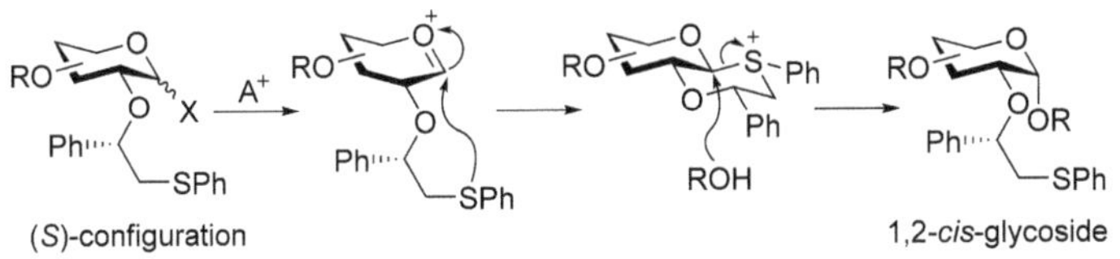

Scheme 45 The 1,2-*cis*-α-glycosylation utilizing an asymmetric auxiliary at O-2.

(Scheme 45).[103] While the concept of neighboring-group participation in the synthesis of 1,2-*trans*-glycosides is highly reliable,[104] the construction of 1,2-*cis*-glycosides such as α-glucosides and -galactosides has been attempted using non-participating protective groups at O-2.[105] However, these donors often give a mixture of both anomers. Thus, Boons and co-workers have developed a method in which (S)-ethoxycarbonylbenzyl[106] and later (1S)-phenyl-2-(phenylsulfanyl)ethyl groups were introduced at O-2 of the donor.[103] When the sulfide of this auxiliary group is added to C-1 of the glycosyl cation, the phenyl group occupies the equatorial position to form a *trans*-decalin system, and the acceptors substitute for the sulfonium group as the leaving group to form 1,2-*cis*-glycosides.

The application of the auxiliary to β-mannosides was hampered by the fact that the C-2 substituent occupies an axial position, where the auxiliary cannot be ring-closed. However, Boltje and co-workers succeeded in enabling the (1R)-phenyl-2-(phenylsulfanyl)ethyl group at O-2 to work as an auxiliary by flipping the mannose ring to $^1C_4$. The 3,6-lactone bridge converted the substituent at O-2 to an equatorial position (Scheme 46, eq. 1).[102] More interestingly, in the presence of a 3,6-lactone, β-stereoselectivity arose even in the absence of an auxiliary group, although the yield was compromised (eq. 2). The authors attributed this to the remote participation of oxygen function at C-4. Indeed, they have detected a 1,4-anhydrosugar, inferred by the strong HMBC correlation among H-1 and C-4, and BnOTf, a by-product generated from the intermediate oxonium species. Alternatively, this β-stereoselectivity might also be explained by a pseudo-axial attack on the oxocarbenium ion in the $^3H_4$ conformation.[94,107]

5.1.2 3,6-O-Xylylene and 3,6-EDB

Yamada and co-workers have reported a β-stereoselective glucosylation using a glucosyl fluoride with *o*-xylyl tethers at the O-3 and -6 positions (Scheme 47).[108] The β-anomer of this donor was fixed to an axial-rich conformation of $^1S_3$. When $SnCl_2$-$AgB(C_6F_5)_4$ was used, the actual activating species was found to be $SnB(C_6F_5)_4Cl$, and both α- and β-glycosides were

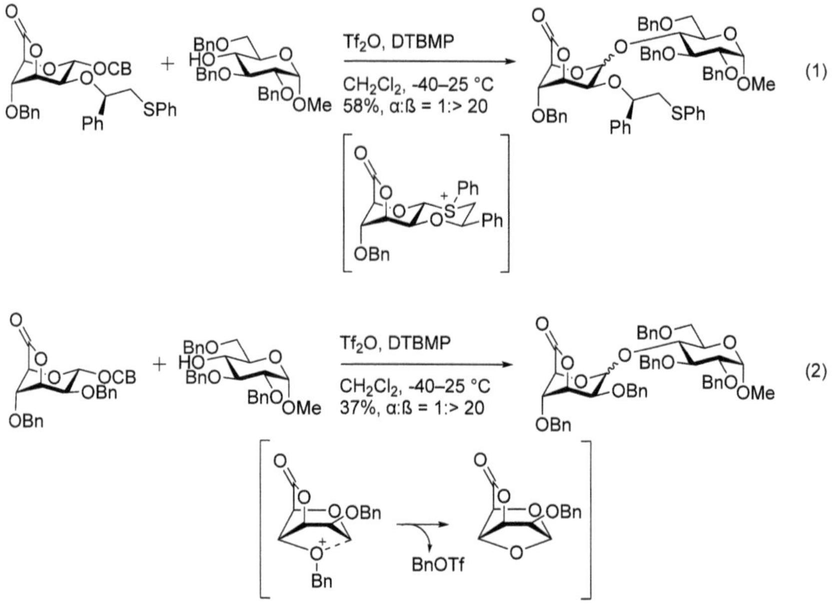

Scheme 46 The 1,2-*cis*-β-glycosylation utilizing an asymmetric auxiliary at O-2 of 3,6-lactone.

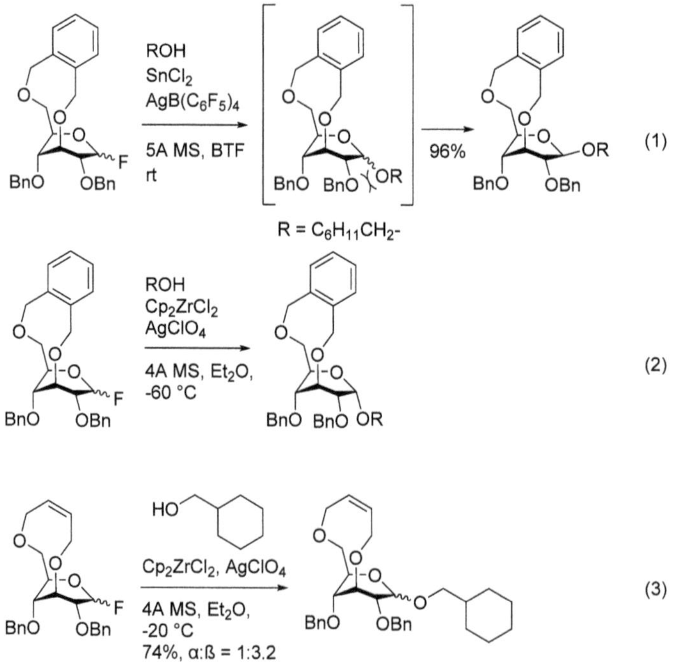

Scheme 47 Either α- or β-stereoselective glucosylations using the 3,6-xylylene tether.

formed. Then, with time, catalyzed by $HB(C_6F_5)_4$ generated in situ, the glycosides were thermodynamically converted to β-anomers, while relaxing the steric repulsion with the C-2 substituent (eq. 1). The authors have also reported that activation of the same donor with Cp_2ZrCl_2-$AgClO_4$ in the presence of 4A molecular sieves in ether kinetically gave preferential access to the α-glycoside (eq. 2).[109] The mechanisms for this was that the benzene ring of xylene masked the β-face of the donor, as revealed by NMR analysis. In addition, they have demonstrated that the donor lacking the benzene ring, which was also in the $^1S_3$ conformation, impaired α-stereoselectivities (eq. 3).

In addition, Yamada and co-workers have performed α-stereoselective glycosylation of perbenzylated glucosyl fluorides whose O-3 and O-6 benzyl groups were tied with ethylene at their ortho positions (Scheme 48).[110] The authors named the tether 3,6-EDB group. In contrast to the robust conformation of ones with the 3,6-o-xylyl group, glucoses with a 3,6-EDB group were supple and showed various conformations depending on the presence or absence of a benzyl group at the O-2 or O-4 position, leaving groups and their stereochemistry at C-1, and the aglycon moiety. Cyclodextrins consisting of three or four glucose residues, CD3 and CD4, which had never been synthesized before, were synthesized for the first time using donors with 3,6-EDB groups. In the structures of the synthesized CD3 and CD4, each sugar residue had a distorted conformation that was not $^4C_1$, and it was assumed that the flexibility of the donors enabled ring-closing glycosylation in tight macrocycles.

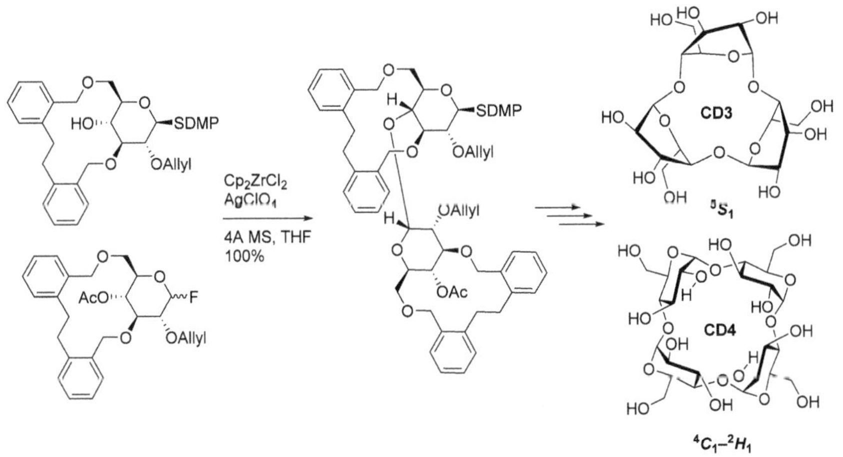

Scheme 48 EDB group enabled syntheses of cyclodextrins consisting of three or four glucose residues.

5.1.3 3,6-Silylene

Pedersen and co-workers report the comprehensively synthesis of phenyl thioglycosides with 3,6-O-silylene bridges in the D-gluco, -manno, and -galacto configurations (Scheme 49).[111] The α-thioglycosides in the gluco configuration were in the $^3S_1$ conformation and the β-anomers in the $B_{1,4}$ conformation. This was probably due to mitigating the flag-pole interaction between the axial substituent at the C-1α and the benzyl group at O-4 in $B_{1,4}$. In fact, in the galacto configuration, where the C-4 substituent was pseudo-equatorial, the thioglycoside was in an almost perfect $B_{1,4}$, independent of the stereochemistry of the anomeric position. In the manno config-uration, the thioglycoside was in $^3S_1$, independent of the stereochemistry of the anomeric position, in order to avoid repulsion between the benzyl group at O-2 and the silylene group. In both cases, the leaving group at the axial or pseudo-axial position was more reactive, favoring E1 elimination by elec-tron donation of the lone pair of the ring oxygen. Furthermore, although the glycosyl cations produced were expected to be in $^3H_4$, α-attacks, which should lead a twisted conformation, were kinetically preferable to β-attacks, which should lead a chair conformation under pseudo-axial attack. This may be due to the advantage of the exo attack against the bulky silylene bridge. In addition, when a glycosyl fluoride was used as the donor and SnCl$_2$-AgB (C$_6$F$_5$)$_4$ as the activator, β-glycosides and α-mannosides were preferentially generated, respectively, by anomerization after glycosylic bond formation.

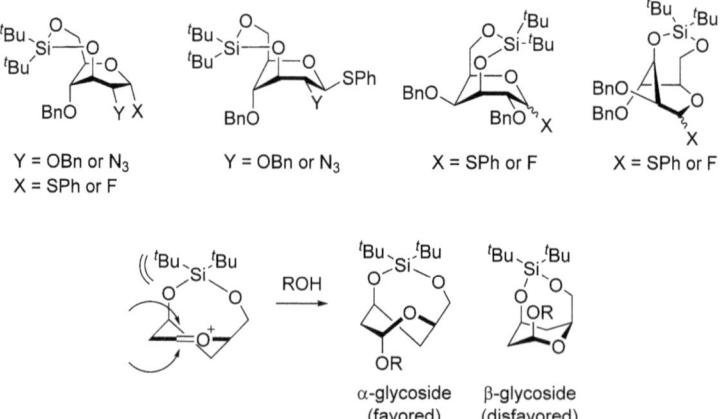

Scheme 49 The conformations of 3,6-silylene donors, and their effects on glycosylations.

5.2 2,6-Tethered donors

5.2.1 2,6-Anhydro-2-thio sugars

Toshima, Tatsuta and colleagues reported a stereoselective reaction using 2,6-anhydro-2-thio sugars as donors (Scheme 50). Activation of the thio-glycoside in the altro configuration with NBS gave the α-glycosides regardless of the stereochemistry of the C-1 position of the donor when the reaction was carried out at low temperature and in a short time (eq. 1).[112] On the other hand, activation of the acetate with TMSOTf stereoselectively gave β-glycosides (eq. 2).[113] The conformation was very rigid in which the anomeric affect did not contribute much to stabilize α-glycosides. The authors assumed that the nucleophilic addition to the glycosyl cation at the β-face was inhibited by the lone pair of the sulfur atom at the bridge, and β-glycosides were thermodynamically formed by steric repulsion with pseudo-axial substit-uents at C-3 or -4. The glycosides can be converted to 2,6-deoxyaltrosides (digitoxosides) via Raney nickel-catalyzed hydrogenation or radical desulfurization (eq. 3).

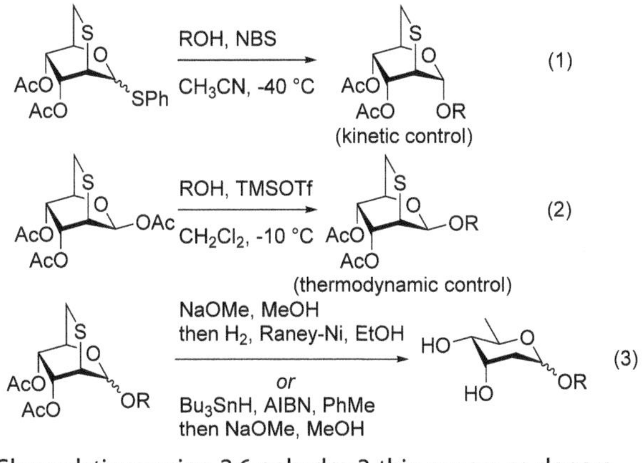

Scheme 50 Glycosylations using 2,6-anhydro-2-thio sugars as donors.

5.2.2 2,6-Lactone donors

Sasaki and co-workers reported that D–manno–configured donors with a 2,6-lactone structure gave β-glycosides in a stereoinversive manner.[114,115] In the $^4C_1$ conformation, which should have been the most extensively studied so far, the lone pair n_o of the ring oxygen overlaps most effectively with the antibonding orbital σ_{C1-X}^* of C-1 and the leaving group X. On the other hand, the 2,6-lactone-containing donors were expected to have a fixed conformation in the range of $^5S_1 \leftrightarrow ^{2,5}B \leftrightarrow ^2S_O$, in which the overlap of n_O and $\sigma_{C1\text{-}X}^*$ should not as effective as in $^4C_1$. Therefore, it was expected that the elimination of the leaving group should slow down, and that S_N2 reaction should become dominant. In fact, β-glycosides could be generated stereoselectively from the donors with a variety of the leaving groups at the α-position. In particular, the reaction proceeded with the highest β-stereoselectivity when trichloroacetoimidate was activated by the combined catalyst of $AuCl_3$ and Schreiner's thiourea (Scheme 51, eq. 1). The stereoselectivity was lost severely in the absence of the 2,6-lactone structure (eq. 2). Since the stereoselectivities were impaired by the use of β-donors and were dependent on the substrate concentration, the authors speculated that the reaction proceeded in an S_N2-like manner. In addition, the C-glycosylation of the 2,6-lactone donor with allylsilane gave exclusively β-glycosides, indicating that the glycosyl cations bearing a 2,6-lactone was β-stereodirecting (eq. 3).

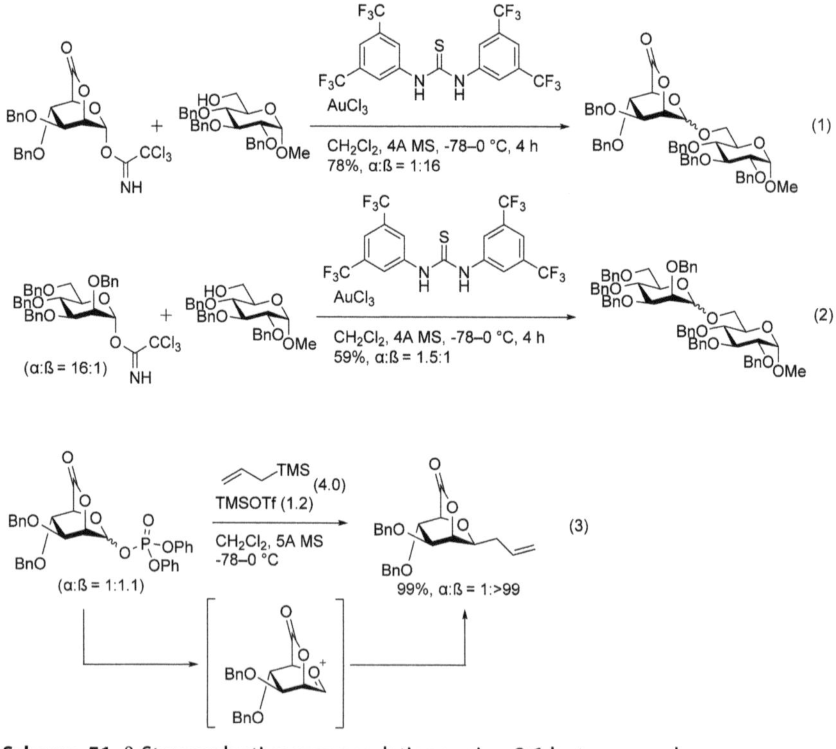

Scheme 51 β-Stereoselective mannosylations using 2,6-lactones as donors.

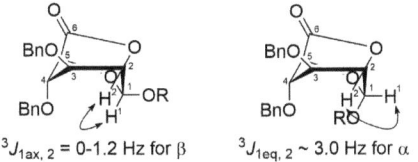

$^3J_{1ax,2}$ = 0–1.2 Hz for β $^3J_{1eq,2}$ ~ 3.0 Hz for α

Fig. 11 Characteristic coupling constants of 2,6-lactone glycosides.

The 2,6-lactone glycoside obtained in the foregoing case was assumed to be in $^5S_1$ based on the long-range coupling between the H-1 and H-5 and the occasional disappearance of the vicial couplings, observed in the ^1H NMR spectra (Fig. 11). For mannosides in $^4C_1$, to determine the stereochemistry of the C-1 position, the coupling constants between the C-1 and H-1 have been used which are quite reliable, but very time–consuming to obtain.[116] On the other hand, for 2,6-lactone glycosides, the coupling constants between the H-1 and H-2 positions, which could be measured in a short time and in tiny quantities, offered a convenient criterion for determining which anomer was generated.[117]

Chai and co-workers have reported that an acyl group at O-4 enhanced β-stereoselectivity in reactions using 2,6-lactone thioglycosides activated by NIS-AgOTf (Scheme 52, eq. 1).[118] The authors claimed that the β-stereoselectivity was due to remote participation, based on trapping of the glycosyl cation with a 4-O-imidoyl group (eq. 2).

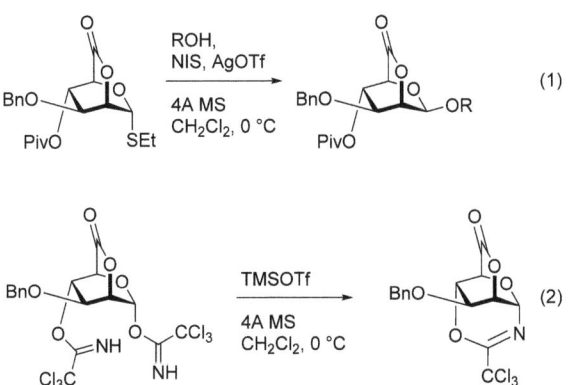

Scheme 52 Remote participation of the acyl groups at O-4 of 2,6-lactones.

6. Conclusions

In this chapter, we have discussed examples of stereoselectivity and reactivity control in chemical glycosylation reactions by controlling conformations of the donors or intermediates. The conformations have been controlled by using steric repulsions of bulky protecting groups, by using cyclic protecting groups, by the oxidation of the 6-position, or by using transannular structures. High stereoselectivities seemed to be achieved by changing the borderline between the S_N2 and S_N1 reactions, or by differentiation between the two faces of the nucleophilic attack or generating new stereorepulsions due to the conformational changes. These may have been the results intended, or they may have been the result of serendipity. Although synthetic carbohydrate chemistry may have been developed based on chemists' experiences and intuitions, we believe that we are making progress toward a deeper and global understanding of this complicated transformation by taking many unique phenomena together. And we hope that glycosylation reactions will progress beyond the trial-and-error stage, and that we will reach an era in which the glycosides can be synthesized at will, as desired.

References

1. Wolfenden, R.; Lu, X.; Young, G. Spontaneous Hydrolysis of Glycosides. *J. Am. Chem. Soc.* **1998**, *120*, 6814–6815.
2. Davies, G. J.; Planas, A.; Rovira, C. Conformational Analyses of the Reaction Coordinate of Glycosidases. *Acc. Chem. Res.* **2012**, *45*, 308–316.
3. Davies, G. J.; Ducros, V. M. A.; Varrot, A.; Zechel, D. L. Mapping the Conformational Itinerary of β-Glycosidases by X-Ray Crystallography. *Biochem. Soc. Trans.* **2003**, *31*, 523–527.
4. Cremer, D.; Pople, J. A. General Definition of Ring Puckering Coordinates. *J. Am. Chem. Soc.* **1975**, *97*, 1354–1358.
5. Davies, G. J.; Mackenzie, L.; Varrot, A.; Dauter, M.; Brzozowski, A. M.; Schülein, M.; Withers, S. G. Snapshots Along an Enzymatic Reaction Coordinate: Analysis of a Retaining β-Glycoside Hydrolase. *Biochemistry* **1998**, *37*, 11707–11713.
6. Ducros, V. M.-A.; Zechel, D. L.; Murshudov, G. N.; Gilbert, H. J.; Szabó, L.; Stoll, D.; Withers, S. G.; Davies, G. J. Substrate Distortion by a β-Mannanase: Snapshots of the Michaelis and Covalent-Intermediate Complexes Suggest a $B_{2,5}$ Conformation for the Transition State. *Angew. Chem., Int. Ed.* **2002**, *41*, 2824–2827.
7. Ando, H.; Komura, N. Chemical Synthesis of Sialoglyco-Architecdtures. *Adv. Carbohydr. Chem. Biochem.* **2022**, *81*. Chapter 3.
8. Nguyen, H.; Zhu, D.; Li, X.; Zhu, J. Stereoselective Construction of β-Mannopyranosides by Anomeric O-Alkylation: Synthesis of the Trisaccharide Core of N-linked Glycans. *Angew. Chem., Int. Ed.* **2016**, *55*, 4767–4771.
9. Yamada, H.; Nakatani, M.; Ikeda, T.; Marumoto, Y. Stable Axial-Rich Conformation of Pyranoses Derived From L-Rhamnose and D-Mannose. *Tetrahedron Lett.* **1999**, *40*, 5573–5576.

10. Yamada, H.; Tanigakiuchi, K.; Nagao, K.; Okajima, K.; Mukae, T. The First Ring Inversion of Pyranoses Induced by Bulky Silyl Protections at the 2- and 3-Positions. *Tetrahedron Lett.* **2004**, *45*, 9207–9209.

11. Abe, H.; Shuto, S.; Tamura, S.; Matsuda, A. An Efficient Method for Preparing Fully O-Silylated Pyranoses Conformationally Restricted in the Unusual $^1C_4$-Form. *Tetrahedron Lett.* **2001**, *42*, 6159–6161.

12. Hosoya, T.; Ohashi, Y.; Matsumoto, T.; Suzuki, K. On the Stereochemistry of Aryl C-Glycosides: Unusual Behavior of Bis-TBDPS-Protected Aryl C-Olivosides. *Tetrahedron Lett.* **1996**, *37*, 663–666.

13. Matsumoto, T.; Katsuki, M.; Jona, H.; Suzuki, K. Synthetic Study Toward Vineomycins. Synthesis of *C*-Aryl Glycoside Sector via Cp_2HfCl_2–$AgClO_4$-Promoted Tactics. *Tetrahedron Lett.* **1989**, *30*, 6185–6188.

14. Matsumoto, T.; Katsuki, M.; Jona, H.; Suzuki, K. Convergent Total Synthesis of Vineomycinone B2 Methyl Ester and its C(12)-Epimer. *J. Am. Chem. Soc.* **1991**, *113*, 6982–6992.

15. Walford, C.; Jackson, R. F. W.; Rees, N. H.; Clegg, W.; Heath, S. L. Reaction of Thiophenol With Glucal Epoxides: X-Ray Structure of 3,4,6-Tri-O-*tert*-butyldimethylsilyl-1-*S*-phenyl-1-thio-α-D-glucopyranoside. *Chem. Commun. (Cambridge)* **1997**, 1855–1856.

16. Tamura, S.; Abe, H.; Matsuda, A.; Shuto, S. Control of α/β Stereoselectivity in Lewis Acid Promoted C-Glycosidations Using a Controlling Anomeric Effect Based on the Conformational Restriction Strategy. *Angew. Chem., Int. Ed.* **2003**, *42*, 1021–1023.

17. Cumpstey, I. On a So-Called "Kinetic Anomeric Effect" in Chemical Glycosylation. *Org. Biomol. Chem.* **2012**, *10*, 2503–2508.

18. van der Vorm, S.; Hansen, T.; Overkleeft, H. S.; van der Marel, G. A.; Codée, J. D. C. The Influence of Acceptor Nucleophilicity on the Glycosylation Reaction Mechanism. *Chem. Sci.* **2017**, *8*, 1867–1875.

19. Okada, Y.; Mukae, T.; Okajima, K.; Taira, M.; Fujita, M.; Yamada, H. Highly β-Selective O-Glucosidation Due to the Restricted Twist-Boat Conformation. *Org. Lett.* **2007**, *9*, 1573–1576.

20. Okada, Y.; Nagata, O.; Taira, M.; Yamada, H. Highly β-Selective and Direct Formation of 2-O-Glycosylated Glucosides by Ring Restriction Into Twist-Boat. *Org. Lett.* **2007**, *9*, 2755–2758.

21. Woods, R. J.; Andrews, C. W.; Bowen, J. P. Molecular Mechanical Investigations of the Properties of Oxocarbenium Ions. 2. Application to Glycoside Hydrolysis. *J. Am. Chem. Soc.* **1992**, *114*, 859–864.

22. Miljković, M.; Yeagley, D.; Deslongchamps, P.; Dory, Y. L. Experimental and Theoretical Evidence of Through-Space Electrostatic Stabilization of the Incipient Oxocarbenium Ion by an Axially Oriented Electronegative Substituent During Glycopyranoside Acetolysis. *J. Org. Chem.* **1997**, *62*, 7597–7604.

23. Jensen, H. H.; Bols, M. Steric Effects Are Not the Cause of the Rate Difference in Hydrolysis of Stereoisomeric Glycosides. *Org. Lett.* **2003**, *5*, 3419–3421.

24. Edward, J. T. Stability of Glycosides to Acid Hydrolysis. A Conformationl Analysis. *Chem. Ind. (London, U. K.)* **1955**, 1102–1104.

25. McDonnell, C.; López, O.; Murphy, P.; Fernández Bolaños, J. G.; Hazell, R.; Bols, M. Conformational Effects on Glycoside Reactivity: Study of the High Reactive Conformer of Glucose. *J. Am. Chem. Soc.* **2004**, *126*, 12374–12385.

26. Zhang, Z.; Ollmann, I. R.; Ye, X.-S.; Wischnat, R.; Baasov, T.; Wong, C.-H. Programmable One-Pot Oligosaccharide Synthesis. *J. Am. Chem. Soc.* **1999**, *121*, 734–753.

27. Jensen, H. H.; Lyngbye, L.; Jensen, A.; Bols, M. Stereoelectronic Substituent Effects in Polyhydroxylated Piperidines and Hexahydropyridazines. *Chem. Eur. J.* **2002**, *8*, 1218–1226.

28. Chamberland, S.; Ziller, J. W.; Woerpel, K. A. Structural Evidence That Alkoxy Substituents Adopt Electronically Preferred Pseudoaxial Orientations in Six-Membered Ring Dioxocarbenium Ions. *J. Am. Chem. Soc.* **2005**, *127*, 5322–5323.

29. Yang, M. T.; Woerpel, K. A. The Effect of Electrostatic Interactions on Conformational Equilibria of Multiply Substituted Tetrahydropyran Oxocarbenium Ions. *J. Org. Chem.* **2009**, *74*, 545–553.

30. Pedersen, C. M.; Nordstrøm, L. U.; Bols, M. "Super Armed" Glycosyl Donors: Conformational Arming of Thioglycosides by Silylation. *J. Am. Chem. Soc.* **2007**, *129*, 9222–9235.

31. Bhat, A. S.; Gervay-Hague, J. Efficient Syntheses of β-Cyanosugars Using Glycosyl Iodides Derived From Per-O-silylated Mono- and Disaccharides. *Org. Lett.* **2001**, *3*, 2081–2084.

32. Crich, D.; Sun, S. Formation of β-Mannopyranosides of Primary Alcohols Using the Sulfoxide Method. *J. Org. Chem.* **1996**, *61*, 4506–4507.

33. Crich, D.; Sun, S. Direct Synthesis of β-Mannopyranosides by the Sulfoxide Method. *J. Org. Chem.* **1997**, *62*, 1198–1199.

34. Crich, D.; Sun, S. Direct Formation of β-Mannopyranosides and Other Hindered Glycosides From Thioglycosides. *J. Am. Chem. Soc.* **1998**, *120*, 435–436.

35. Crich, D.; Smith, M. 1-Benzenesulfinyl Piperidine/Trifluoromethanesulfonic Anhydride: A Potent Combination of Shelf-Stable Reagents for the Low-Temperature Conversion of Thioglycosides to Glycosyl Triflates and for the Formation of Diverse Glycosidic Linkages. *J. Am. Chem. Soc.* **2001**, *123*, 9015–9020.

36. Kim, K. S.; Kim, J. H.; Lee, Y. J.; Lee, Y. J.; Park, J. 2-(Hydroxycarbonyl)benzyl Glycosides: A Novel Type of Glycosyl Donors for Highly Efficient β-Mannopyranosylation and Oligosaccharide Synthesis by Latent-Active Glycosylation. *J. Am. Chem. Soc.* **2001**, *123*, 8477–8481.

37. Kim, K. S.; Lee, Y. J.; Kim, H. Y.; Kang, S. S.; Kwon, S. Y. Glycosylation with Glycosyl enzyl Phthalates as a New Type of Glycosyl Donor. *Org. Biomol. Chem.* **2004**, *2*, 2408–2410.

38. Baek, J. Y.; Choi, T. J.; Jeon, H. B.; Kim, K. S. A Highly Reactive and Stereoselective β-Mannopyranosylation System: Mannosyl 4-Pentenoate/PhSeOTf. *Angew. Chem., Int. Ed.* **2006**, *45*, 7436–7440.

39. Codée, J. D. C.; Hossain, L. H.; Seeberger, P. H. Efficient Installation of β-Mannosides Using a Dehydrative Coupling Strategy. *Org. Lett.* **2005**, *7*, 3251–3254.

40. Weingart, R.; Schmidt, R. R. Can Preferential β-Mannopyranoside Formation With 4,6-O-Benzylidene-Protected Mannopyranosyl Sulfoxides be Reached with Trichloroacetimidates? *Tetrahedron Lett.* **2000**, *41*, 8753–8758.

41. Tsuda, T.; Sato, S.; Nakamura, S.; Hashimoto, S. Direct and stereoselective Construction of β-Mannosidic Linkages Capitalizing on 4,6-O-Benzylidene-Protected D-mannopyranosyl Diethyl Phosphite. *Heterocycles* **2003**, *59*, 509–515.

42. Tanaka, S.-i.; Takashina, M.; Tokimoto, H.; Fujimoto, Y.; Tanaka, K.; Fukase, K. Highly β-Selective Mannosylation towards Manβ1-4GlcNAc Synthesis: TMSB(C$_6$F$_5$)$_4$ as a Lewis Acid/Cation Trap Catalyst. *Synlett* **2005**, 2325–2328.

43. Sun, P.; Wang, P.; Zhang, Y.; Zhang, X.; Wang, C.; Liu, S.; Lu, J.; Li, M. Construction of β-Mannosidic Bonds via Gold(I)-Catalyzed Glycosylations with Mannopyranosyl *ortho*-Hexynylbenzoates and Its Application in Synthesis of Acremomannolipin A. *J. Org. Chem.* **2015**, *80*, 4164–4175.

44. Crich, D.; Smith, M. Solid-Phase Synthesis of β-Mannosides. *J. Am. Chem. Soc.* **2002**, *124*, 8867–8869.

45. Crich, D.; de la Mora, M.; Vinod, A. U. Influence of the 4,6-O-Benzylidene, 4,6-O-Phenylboronate, and 4,6-O-Polystyrylboronate Protecting Groups on the Stereochemical Outcome of Thioglycoside-Based Glycosylations Mediated by

1-Benzenesulfinyl Piperidine/Triflic Anhydride and N-Iodosuccinimide/ Trimethylsilyl Triflate. *J. Org. Chem.* **2003**, *68*, 8142–8148.

46. Crich, D.; Li, L. Stereocontrolled Synthesis of D- and L-β-Rhamnopyranosides with 4-O-6-S-α-Cyanobenzylidene-Protected 6-Thiorhamnopyranosyl Thioglycosides. *J. Org. Chem.* **2009**, *74*, 773–781.

47. Crich, D.; Banerjee, A. Stereocontrolled Synthesis of the D- and L-*glycero*-β-D-*manno*-Heptopyranosides and Their 6-Deoxy Analogues. Synthesis of Methyl α-L-*rhamno*-Pyranosyl-(1→3)-D-*glycero*-β-D-*manno*-heptopyranosyl-(1→3)-6-deoxy-*glycero*-β-D-*manno*-heptopyranosyl-(1→4)-α-L-*rhamno*-pyranoside, a Tetrasaccharide Subunit of the Lipopolysaccharide From Plesimonas shigelloides. *J. Am. Chem. Soc.* **2006**, *128*, 8078–8086.

48. Heuckendorff, M.; Bendix, J.; Pedersen, C. M.; Bols, M. β-Selective Mannosylation With a 4,6-Silylene-Tethered Thiomannosyl Donor. *Org. Lett.* **2014**, *16*, 1116–1119.

49. Fraser-Reid, B.; Wu, Z.; Andrews, C. W.; Skowronski, E.; Bowen, J. P. Torsional Effects in Glycoside Reactivity: Saccharide Couplings Mediated by Acetal Protecting Groups. *J. Am. Chem. Soc.* **1991**, *113*, 1434–1435.

50. Ley, S. V.; Priepke, H. W. M. A Facile One-Pot Synthesis of a Trisaccharide Unit From the Common Polysaccharide Antigen of Group B Streptococci Using Cyclohexane-1, 2-diacetal (CDA) Protected Rhamnosides. *Angew. Chem., Int. Ed. Engl.* **1994**, *33*, 2292–2294.

51. Jensen, H. H.; Nordstrøm, L. U.; Bols, M. The Disarming Effect of the 4,6-Acetal Group on Glycoside Reactivity: Torsional or Electronic? *J. Am. Chem. Soc.* **2004**, *126*, 9205–9213.

52. Moumé-Pymbock, M.; Furukawa, T.; Mondal, S.; Crich, D. Probing the Influence of a 4,6-O-Acetal on the Reactivity of Galactopyranosyl Donors: Verification of the Disarming Influence of the trans–gauche Conformation of C5–C6 Bonds. *J. Am. Chem. Soc.* **2013**, *135*, 14249–14255.

53. Frihed, T. G.; Walvoort, M. T. C.; Codée, J. D. C.; van der Marel, G. A.; Bols, M.; Pedersen, C. M. Influence of O6 in Mannosylations Using Benzylidene Protected Donors: Stereoelectronic or Conformational Effects? *J. Org. Chem.* **2013**, *78*, 2191–2205.

54. Hosoya, T.; Kosma, P.; Rosenau, T. Theoretical Study on the Effects of a 4,6-O-Diacetal Protecting Group on the Stability of Ion Pairs From D-mannopyranosyl and D-Glucopyranosyl Triflates. *Carbohydr. Res.* **2015**, *411*, 64–69.

55. Li, Z. Computational Study of the Influence of Cyclic Protecting Groups in Stereoselectivity of Glycosylation Reactions. *Carbohydr. Res.* **2010**, *345*, 1952–1957.

56. Hosoya, T.; Kosma, P.; Rosenau, T. Contact Ion Pairs and Solvent-Separated Ion Pairs From D-Mannopyranosyl and D-Glucopyranosyl Triflates. *Carbohydr. Res.* **2015**, *401*, 127–131.

57. Whitfield, D. M. DFT Studies of the Ionization of α- and β-Glycopyranosyl Donors. *Carbohydr. Res.* **2007**, *342*, 1726–1740.

58. Huang, M.; Garrett, G. E.; Birlirakis, N.; Bohé, L.; Pratt, D. A.; Crich, D. Dissecting the Mechanisms of a Class of Chemical Glycosylation using Primary ^{13}C Kinetic Isotope Effects. *Nat. Chem.* **2012**, *4*, 663–667.

59. Bousquet, E.; Khitri, M.; Lay, L.; Nicotra, F.; Panza, L.; Russo, G. Capsular Polysaccharide of *Streptococcus pneumoniae* Type 19F: Synthesis of the Repeating Unit. *Carbohydr. Res.* **1998**, *311*, 171–181.

60. Crich, D.; Cai, W. Chemistry of 4,6-O-Benzylidene-D-glycopyranosyl Triflates: Contrasting Behavior between the Gluco and Manno Series. *J. Org. Chem.* **1999**, *64*, 4926–4930.

61. Heuckendorff, M.; Jensen, H. H. On the Gluco/Manno Paradox: Practical α-Glucosylations by NIS/TfOH Activation of 4,6-O-Tethered Thioglucoside Donors. *Eur. J. Org. Chem.* **2016**, *2016*, 5136–5145.

62. van der Vorm, S.; Overkleeft, H. S.; van der Marel, G. A.; Codée, J. D. C. Stereoselectivity of Conformationally Restricted Glucosazide Donors. *J. Org. Chem.* **2017**, *82*, 4793–4811.

63. Izumi, M.; Shen, G.-J.; Wacowich-Sgarbi, S.; Nakatani, T.; Plettenburg, O.; Wong, C.-H. Microbial Glycosyltransferases for Carbohydrate Synthesis: α-2,3-Sialyltransferase from *Neisseria gonorrheae*. *J. Am. Chem. Soc.* **2001**, *123*, 10909–10918.

64. Imamura, A.; Ando, H.; Korogi, S.; Tanabe, G.; Muraoka, O.; Ishida, H.; Kiso, M. Di-*tert*-butylsilylene (DTBS) Group-Directed α-Selective Galactosylation Unaffected by C-2 Participating Functionalities. *Tetrahedron Lett.* **2003**, *44*, 6725–6728.

65. Imamura, A.; Kimura, A.; Ando, H.; Ishida, H.; Kiso, M. Extended Applications of Di-*tert*-butylsilylene-Directed α-Predominant Galactosylation Compatible with C2-Participating Groups Toward the Assembly of Various Glycosides. *Chem. Eur. J.* **2006**, *12*, 8862–8870.

66. Imamura, A.; Matsuzawa, N.; Sakai, S.; Udagawa, T.; Nakashima, S.; Ando, H.; Ishida, H.; Kiso, M. The Origin of High Stereoselectivity in Di-tert-butylsilylene-Directed α-Galactosylation. *J. Org. Chem.* **2016**, *81*, 9086–9104.

67. Yasomanee, J. P.; Demchenko, A. V. Effect of Remote Picolinyl and Picoloyl Substituents on the Stereoselectivity of Chemical Glycosylation. *J. Am. Chem. Soc.* **2012**, *134*, 20097–20102.

68. Pistorio, S. G.; Yasomanee, J. P.; Demchenko, A. V. Hydrogen-Bond–Mediated Aglycone Delivery: Focus on β-Mannosylation. *Org. Lett.* **2014**, *16*, 716–719.

69. Alex, C.; Visansirikul, S.; Demchenko, A. V. A Versatile Approach to the Synthesis of Mannosamine Glycosides. *Org. Biomol. Chem.* **2020**, *18*, 6682–6695.

70. Crich, D.; Cai, W.; Dai, Z. Highly Diastereoselective α-Mannopyranosylation in the Absence of Participating Protecting Groups. *J. Org. Chem.* **2000**, *65*, 1291–1297.

71. Crich, D.; Yao, Q. Benzylidene Acetal Fragmentation Route to 6-Deoxy Sugars: Direct Reductive Cleavage in the Presence of Ether Protecting Groups, Permitting the Efficient, Highly Stereocontrolled Synthesis of β-D-Rhamnosides from D-Mannosyl Glycosyl Donors. Total Synthesis of α-D-Gal-(1→3)-α-D-Rha-(1→3)-β-D-Rha-(1→4)-β-D-Glu-OMe, the Repeating Unit of the Antigenic Lipopolysaccharide From *Escherichia hermannii* ATCC 33650 and 33652. *J. Am. Chem. Soc.* **2004**, *126*, 8232–8236.

72. Baek, J. Y.; Lee, B.-Y.; Jo, M. G.; Kim, K. S. β-Directing Effect of Electron-Withdrawing Groups at O-3, O-4, and O-6 Positions and α-Directing Effect by Remote Participation of 3-O-Acyl and 6-O-Acetyl Groups of Donors in Mannopyranosylations. *J. Am. Chem. Soc.* **2009**, *131*, 17705–17713.

73. Crich, D.; Hu, T.; Cai, F. Does Neighboring Group Participation by Non-Vicinal Esters Play a Role in Glycosylation Reactions? Effective Probes for the Detection of Bridging Intermediates. *J. Org. Chem.* **2008**, *73*, 8942–8953.

74. Codée, J. D. C.; de Jong, A. R.; Dinkelaar, J.; Overkleeft, H. S.; van der Marel, G. A. Stereoselectivity of Glycosylations of Conformationally Restricted Mannuronate Esters. *Tetrahedron* **2009**, *65*, 3780–3788.

75. Iversen, T.; Bundle, D. R. A New and Efficient Synthesis of β-L-Rhamnopyranosides. *Carbohydr. Res.* **1980**, *84*, c13–c15.

76. Backinowsky, L. V.; Balan, N. F.; Shashkov, A. S.; Kochetkov, N. K. Synthesis and ^{13}C-N.M.R. Spectra of β-L-Rhamnopyranosides. *Carbohydr. Res.* **1980**, *84*, 225–235.

77. Crich, D.; Vinod, A. U.; Picione, J. The 3,4-O-Carbonate Protecting Group as a β-Directing Group in Rhamnopyranosylation in Both Homogeneous and Heterogeneous Glycosylations As Compared to the Chameleon-Like 2,3-O-Carbonates. *J. Org. Chem.* **2003**, *68*, 8453–8458.

78. Crich, D.; Subramanian, V.; Hutton, T. K. β-Selective Glucosylation in the Absence of Neighboring Group Participation: Influence of the 3,4-O-Bisacetal Protecting System. *Tetrahedron* **2007**, *63*, 5042–5049.

79. Andrews, C. W.; Rodebaugh, R.; Fraser-Reid, B. A Solvation-Assisted Model for Estimating Anomeric Reactivity. Predicted Versus Observed Trends in Hydrolysis of *n*-Pentenyl Glycosides1. *J. Org. Chem.* **1996**, *61*, 5280–5289.

80. Yagami, N.; Tamai, H.; Udagawa, T.; Ueki, A.; Konishi, M.; Imamura, A.; Ishida, H.; Kiso, M.; Ando, H. A 1,2-trans-Selective Glycosyl Donor Bearing Cyclic Protection at the C-2 and C-3 Hydroxy Groups. *Eur. J. Org. Chem.* **2017**, *2017*, 4778–4785.

81. Benakli, K.; Zha, C.; Kerns, R. J. Oxazolidinone Protected 2-Amino-2-deoxy-D-glucose Derivatives as Versatile Intermediates in Stereoselective Oligosaccharide Synthesis and the Formation of α-Linked Glycosides. *J. Am. Chem. Soc.* **2001**, *123*, 9461–9462.

82. Crich, D.; Vinod, A. U. Oxazolidinone Protection of *N*-Acetylglucosamine Confers High Reactivity on the 4-Hydroxy Group in Glycosylation. *Org. Lett.* **2003**, *5*, 1297–1300.

83. Zhu, T.; Boons, G.-J. Thioglycosides Protected as *trans*-2,3-Cyclic Carbonates in Chemoselective Glycosylations. *Org. Lett.* **2001**, *3*, 4201–4203.

84. Boysen, M.; Gemma, E.; Lahmann, M.; Oscarson, S. Ethyl 2-Acetamido-4,6-di-O-benzyl-2,3-N,O-carbonyl-2-deoxy-1-thio-β-D-glycopyranoside as a Versatile GlcNAc Donor. *Chem. Commun. (Cambridge)* **2005**, 3044–3046.

85. Olsson, J. D. M.; Eriksson, L.; Lahmann, M.; Oscarson, S. Investigations of Glycosylation Reactions With 2-N-Acetyl-2-N,3-O-oxazolidinone-Protected Glucosamine Donors. *J. Org. Chem.* **2008**, *73*, 7181–7188.

86. Satoh, H.; Manabe, S.; Ito, Y.; Lüthi, H. P.; Laino, T.; Hutter, J. Endocyclic Cleavage in Glycosides With 2,3-trans Cyclic Protecting Groups. *J. Am. Chem. Soc.* **2011**, *133*, 5610–5619.

87. Geng, Y.; Qin, Q.; Ye, X.-S. Lewis Acids as α-Directing Additives in Glycosylations by Using 2,3-O-Carbonate-Protected Glucose and Galactose Thioglycoside Donors Based on Preactivation Protocol. *J. Org. Chem.* **2012**, *77*, 5255–5270. https://doi.org/10.1021/jo3002084.

88. Lu, Y.-S.; Li, Q.; Zhang, L.-H.; et al. Highly Direct α-Selective Glycosylations of 3,4-O-Carbonate-Protected 2-Deoxy- and 2,6-Dideoxythioglycosides by Preactivation Protocol. *Org. Lett.* **2008**, *10*, 3445–3448.

89. Toshima, K.; Nozaki, Y.; Tatsuta, K. Highly Stereoselective Glycosylation by Conformational Assistance of Glycosyl Donor. *Tetrahedron Lett.* **1991**, *32*, 6887–6890.

90. Durham, T. B.; Roush, W. R. Stereoselective Synthesis of 2-Deoxy-β-galactosides via 2-Deoxy-2-bromo- and 2-Deoxy-2-iodo-galactopyranosyl Donors. *Org. Lett.* **2003**, *5*, 1871–1874.

91. Martin, A.; Arda, A.; Désiré, J.; Martin-Mingot, A.; Probst, N.; Sinaÿ, P.; Jiménez-Barbero, J.; Thibaudeau, S.; Blériot, Y. Catching Elusive Glycosyl Cations in a Condensed Phase With HF/SbF5 Superacid. *Nat. Chem.* **2016**, *8*, 186–191.

92. Balmond, E. I.; Benito-Alifonso, D.; Coe, D. M.; Alder, R. W.; McGarrigle, E. M.; Galan, M. C. A 3,4-*trans*-Fused Cyclic Protecting Group Facilitates α-Selective Catalytic Synthesis of 2-Deoxyglycosides. *Angew. Chem., Int. Ed.* **2014**, *53*, 8190–8194.

93. van den Bos, L. J.; Dinkelaar, J.; Overkleeft, H. S.; van der Marel, G. A. Stereocontrolled Synthesis of β-D-Mannuronic Acid Esters: Synthesis of an Alginate Trisaccharide. *J. Am. Chem. Soc.* **2006**, *128*, 13066–13067.

94. Codée, J. D. C.; van den Bos, L. J.; de Jong, A.-R.; Dinkelaar, J.; Lodder, G.; Overkleeft, H. S.; van der Marel, G. A. The Stereodirecting Effect of the Glycosyl C5-Carboxylate Ester: Stereoselective Synthesis of β-Mannuronic Acid Alginates. *J. Org. Chem.* **2009**, *74*, 38–47.

95. Walvoort, M. T. C.; Lodder, G.; Mazurek, J.; Overkleeft, H. S.; Codée, J. D. C.; van der Marel, G. A. Equatorial Anomeric Triflates From Mannuronic Acid Esters. *J. Am. Chem. Soc.* **2009**, *131*, 12080–12081.

96. Elferink, H.; Severijnen, M. E.; Martens, J.; Mensink, R. A.; Berden, G.; Oomens, J.; Rutjes, F. P. J. T.; Rijs, A. M.; Boltje, T. J. Direct Experimental Characterization of Glycosyl Cations by Infrared Ion Spectroscopy. *J. Am. Chem. Soc.* **2018**, *140*, 6034–6038.

97. Nukada, T.; Bérces, A.; Wang, L.; Zgierski, M. Z.; Whitfield, D. M. The Two-Conformer Hypothesis: 2,3,4,6-*tetra*-O-Methyl-mannopyranosyl and -glucopyranosyl Oxacarbenium Ions. *Carbohydr. Res.* **2005**, *340*, 841–852.

98. Alex, C.; Visansirikul, S.; Demchenko, A. V. A Versatile Approach to the Synthesis of Glycans Containing Mannuronic Acid Residues. *Org. Biomol. Chem.* **2021**, *19*, 2731–2743.

99. Mannino, M. P.; Demchenko, A. V. Synthesis of β-Glucosides With 3-O-Picoloyl-Protected Glycosyl Donors in the Presence of Excess Triflic Acid: A Mechanistic Study. *Chem. Eur. J.* **2020**, *26*, 2927–2937.

100. van den Bos, L. J.; Litjens, R. E. J. N.; van den Berg, R. J. B. H. N.; Overkleeft, H. S.; van der Marel, G. A. Preparation of 1-Thio Uronic Acid Lactones and Their Use in Oligosaccharide Synthesis. *Org. Lett.* **2005**, *7*, 2007–2010.

101. Christina, A. E.; van den Bos, L. J.; Overkleeft, H. S.; van der Marel, G. A.; Codée, J. D. C. Galacturonic Acid Lactones in the Synthesis of All Trisaccharide Repeating Units of the Zwitterionic Polysaccharide Sp1. *J. Org. Chem.* **2011**, *76*, 1692–1706.

102. Elferink, H.; Mensink, R. A.; White, P. B.; Boltje, T. J. Stereoselective β-Mannosylation by Neighboring-Group Participation. *Angew. Chem., Int. Ed.* **2016**, *55*, 11217–11220.

103. Kim, J.-H.; Yang, H.; Park, J.; Boons, G.-J. A General Strategy for Stereoselective Glycosylations. *J. Am. Chem. Soc.* **2005**, *127*, 12090–12097.

104. Hettikankanamalage, A. A.; Lassfolk, R.; Ekholm, F. S.; Leino, R.; Crich, D. Mechanisms of Stereodirecting Participation and Ester Migration From Near and Far in Glycosylation and Related Reactions. *Chem. Rev.* **2020**, *120*, 7104–7151.

105. Demchenko, A. V. Stereoselective Chemical 1,2-cis O-Glycosylation: From 'Sugar Ray' to Modern Techniques of the 21st Century. *Synlett* **2003**, *2003*, 1225–1240.

106. Kim, J.-H.; Yang, H.; Boons, G.-J. Stereoselective Glycosylation Reactions With Chiral Auxiliaries. *Angew. Chem., Int. Ed.* **2005**, *44*, 947–949.

107. Ayala, L.; Lucero, C. G.; Romero, J. A. C.; Tabacco, S. A.; Woerpel, K. A. Stereochemistry of Nucleophilic Substitution Reactions Depending Upon Substituent: Evidence for Electrostatic Stabilization of Pseudoaxial Conformers of Oxocarbenium Ions by Heteroatom Substituents. *J. Am. Chem. Soc.* **2003**, *125*, 15521–15528.

108. Okada, Y.; Asakura, N.; Bando, M.; Ashikaga, Y.; Yamada, H. Completely β-Selective Glycosylation Using 3,6-O-(*o*-Xylylene)-Bridged Axial-Rich Glucosyl Fluoride. *J. Am. Chem. Soc.* **2012**, *134*, 6940–6943.

109. Motoyama, A.; Arai, T.; Ikeuchi, K.; Aki, K.; Wakamori, S.; Yamada, H. α-Selective Glycosylation of 3,6-O-*o*-Xylylene-Bridged Glucosyl Fluoride. *Synthesis* **2018**, *50*, 282–294.

110. Ikuta, D.; Hirata, Y.; Wakamori, S.; Shimada, H.; Tomabechi, Y.; Kawasaki, Y.; Ikeuchi, K.; Hagimori, T.; Matsumoto, S.; Yamada, H. Conformationally Supple Glucose Monomers Enable Synthesis of the Smallest Cyclodextrins. *Science* **2019**, *364*, 674–677.

111. Heuckendorff, M.; Pedersen, C. M.; Bols, M. Conformationally Armed 3,6-Tethered Glycosyl Donors: Synthesis, Conformation, Reactivity, and Selectivity. *J. Org. Chem.* **2013**, *78*, 7234–7248.

112. Toshima, K.; Mukaiyama, S.; Ishiyama, T.; Tatsuta, K. The Use of 2,6-Anhydro-2-Thio Sugar for a Highly Stereocontrolled Glycosylation: A Novel Strategy for Synthesis of 2,6-Dideoxy-α-glycosides. *Tetrahedron Lett.* **1990**, *31*, 3339–3342.

113. Toshima, K.; Mukaiyama, S.; Nozaki, Y.; Inokuchi, H.; Nakata, M.; Tatsuta, K. Novel Glycosidation Method Using 2,6-Anhydro-2-thio Sugars for Stereocontrolled Synthesis of 2,6-Dideoxy-α- and β-glycosides. *J. Am. Chem. Soc.* **1994**, *116*, 9042–9051.
114. Sasaki, K.; Hashimoto, Y. 2,6-Lactones as a New Entry in Stereoselective Glycosylations. *Synlett* **2017**, 1121–1126.
115. Hashimoto, Y.; Tanikawa, S.; Saito, R.; Sasaki, K. β-Stereoselective Mannosylation Using 2,6-Lactones. *J. Am. Chem. Soc.* **2016**, *138*, 14840–14843.
116. Bock, K.; Pedersen, C. A Study of ^{13}CH Coupling Constants in Hexopyranoses. *J. Chem. Soc., Perkin Trans. 2* **1974**, 293–297.
117. Tohda, K.; Saito, M.; Sakai, H.; Ishikura, D.; Saito, R.; Sasaki, K. NMR Characterization of α- and β-Mannopyranurono-2,6-lactones. *Tetrahedron* **2018**, *74*, 5481–5485.
118. Xu, H.; Chen, L.; Zhang, Q.; Feng, Y.; Zu, Y.; Chai, Y. Stereoselective β-Mannosylation With 2,6-Lactone-Bridged Thiomannosyl Donor by Remote Acyl Group Participation. *Chem. Asian J.* **2019**, *14*, 1424–1428.

Author index

Subject index

Note: Page numbers followed by "*f*" indicate figures and "*s*" indicate schemes.

Lightning Source UK Ltd.
Milton Keynes UK
UKHW020104241222
414375UK00001B/40

9 780323 985970